Contents

Introduction	4
Classification	5
Field hints	8
Where to look	8
How to use this book	9
Identification in the field	10
Snake bites: dos and don'ts	10
Species accounts	
Blind snakes	11–13
Pythons	14–21
File snakes	22–23
Colubrid snakes	24–29
Elapid snakes	30–63
Sea-snakes	64–69
Sea kraits	70
Crocodiles	71–72
Marine turtles	73–75
Freshwater turtles	76–79
Geckos	80–89
Pygopodids	90–95
Monitors or Goannas	96–102
Dragons	103–112
Skinks	113–139
Glossary	140
Further reading	141
Index	142

INTRODUCTION

With over 760 species known, Australia has a reptile fauna that, in terms of numbers and diversity, provides ample opportunity for observation by those with an interest in these animals.

All reptiles have an impermeable, scaly skin which helps reduce loss of water from the body. This, together with their ability to convert soluble body wastes, primarily urea, into solid uric acid provides them with an efficient body water-conservation system, enabling terrestrial reptiles to colonize arid areas successfully. Aquatic reptiles, of course, do not have the same need to conserve body water and most of them pass a liquid urine.

Having an inelastic body covering means that, as it grows, a reptile must shed its skin as a new one forms beneath it. This process is called sloughing. Prior to sloughing, the skin may become milky, and this is particularly noticeable in the eyes of a snake which turn quite opaque. Sloughing is usually achieved piecemeal in lizards and it is not unusual to find individuals with pieces of old skin still adhering to different parts of the body. In snakes the skin is usually shed in one piece by the old skin being turned inside out. A snake achieves this by rubbing its snout against a fixed object to start the peeling process. The old skin rolls back over the head and, as it catches against other objects, it is drawn back over the snake, which, by moving around, pulls itself out of the old skin.

Many lizards, particularly geckos, pygopodids and skinks, have the ability to cast off all or part of the tail to distract the attention of a potential predator, a process known as autotomy. Another tail eventually grows to replace the lost portion but is usually quite different in colour or texture and lacks the vertebral bones of the original. (The Keelback Snake also has this ability.) Occasionally, lizards are found with two or three tails. This can happen when the original tail is partially broken but does not break off; it heals, but a new tail also grows from the place where the partial break occurred.

Reptiles are ectothermic animals which means that they must use an external source of heat to raise the body temperature to the preferred level for normal activities. This is usually achieved by basking in direct sunlight or by absorbing heat from a warmed surface. To lower its body temperature, or to maintain it at optimum level, a reptile moves into the shade or alters its body posture to reduce heat uptake. This method of heat regulation, while ideal in tropical and warm parts of the world, puts reptiles at a disadvantage in cooler environments. In these regions, for at least part of the year, daytime temperatures are insufficient for the animal to raise its temperature to that required for any activity. In these circumstances reptiles typically enter a period of hibernation or torpor in a suitable shelter site.

Most reptiles lay eggs, which are deposited in crevices, burrows, cavities beneath rocks and the like. In Australia only pythons and crocodiles demonstrate any parental care of the eggs after they have been laid. In some species, however, the female retains the eggs within her body until they hatch. In others the embryos develop a placental attachment to the oviduct. These latter two modes are often found in reptiles inhabiting cooler regions.

Generally, reptiles are carnivorous and consume a wide variety of prey. Some snakes in other parts of the world feed on slugs, snails, centipedes or insects but there is very little evidence to demonstrate that Australian snakes eat invertebrates, with the obvious exception of the blind snakes of the family Typhlopidae which feed on ant and termite pupae, larvae and eggs. Snakes of our other six

A Photographic Guide to SNAKES & OTHER REPTILES OF AUSTRALIA

Gerry Swan
The Australian Museum

First published by
New Holland (Publishers) Ltd
London • Cape Town • Sydney • Singapore

Copyright © 1995 New Holland (Publishers) Ltd
Copyright © in text The Australian Museum Trust
Copyright © in photographs as indicated

All rights reserved. No part of this publication may be reproduced, stored in a retrieval system or transmitted, in any form or by any means, electronic, mechanical, photocopying, recording or otherwise, without the prior written permission of the publishers and copyright holders.

ISBN 185368 585 2

New Holland Publishers (Australia)
An imprint of National Book Distributors Pty Ltd
3/2 Aquatic Drive
Frenchs Forest
NSW 2086
Australia

Editor: John Comrie-Greig
Typeset by EPS&M
Reproduction by Hirt & Carter
Printed and bound by Tien Wah Press (Pte) Ltd

Front cover: H. & J. Beste
Back cover: G.I. Little
Spine: D. & V. Bladgen
Title page: R. & D. Keller

families feed mainly on lizards, frogs, other snakes, reptile eggs, birds or mammals, all of which are swallowed whole. Snakes can go for some weeks or more without eating. Most lizards are carnivorous but many are also partly herbivorous. Some turtles elsewhere in the world are entirely herbivorous but all Australian species are predominantly carnivorous, although some eat a little aquatic vegetation and adult green turtles prefer marine grasses.

Classification

There are four main lineages or orders of reptiles still surviving today, of which three are represented in Australia.

The order **Crocodilia** comprises the alligators and crocodiles. It is represented in Australia by two species of crocodiles.

The order **Testudines** comprises the tortoises, terrapins and turtles. It is represented in Australia by freshwater and marine turtles, but not by land tortoises.

The order **Squamata** comprises the snakes, lizards and amphisbaenids (often incorrectly called 'worm lizards'). This order is well represented in Australia by snakes and lizards, but there are no amphisbaenids.

(The order **Rhynchocephalia** has only one surviving species, the Tuatara of New Zealand.)

ORDER CROCODILIA
Of the three extant families, only one is represented in Australia.

FAMILY CROCODYLIDAE (PAGES 71–72)
The crocodiles occurring in Australia share the generalized features of their kin worldwide, with a long, lizard-like body covered in bony plates overlain by thick horny scales. They are carnivorous and aquatic, with short powerful limbs and webbed feet. The nostrils are valvular and situated on a raised area near the tip of the long snout. There are no alligators native to Australia.

ORDER TESTUDINES

This order contains 11 families of tortoises and turtles of which four are found in Australia. The body of a typical turtle is completely enclosed in a shell which is made up of an upper half, the carapace, and a lower half to protect the underside, the plastron. The shell consists of fused bony plates that are covered (in most species) with horny plates or shields.

FAMILY CHELONIIDAE (PAGES 73–75)
These are marine turtles with limbs that have been modified to form paddle-like flippers, each with one or two claws. The head cannot be withdrawn into the shell and the carapace and plastron are covered by horny plates or shields. The females must come ashore to lay their eggs.

FAMILY DERMOCHELYIDAE (PAGE 75)
There is only one species in this family, the Leathery Turtle, which has a leathery skin covering its bony shell rather than horny plates. It is marine and the paddle-shaped limbs have no claws. The head cannot be pulled back into the shell. The females must come ashore to lay their eggs as with all other marine turtle species.

FAMILY CARETTOCHELYDIDAE (PAGE 76)
The Pitted-shelled Turtle is the only species in this family. It inhabits fresh water and has its carapace covered in a soft skin rather than the hard keratin or 'horn' of other species. The head cannot be pulled back into the shell and the limbs are paddle-like with two claws to each. The eggs are laid on land.

FAMILY CHELIDAE (PAGES 77–79)
These freshwater turtles have a hard-shelled carapace and plastron. They have clawed, webbed feet and the head can be pulled back under the edge of the carapace by folding the neck in a horizontal S-shape. All species lay their eggs on land.

ORDER SQUAMATA

SUBORDER SERPENTES
Snakes are represented worldwide by some 11 families, of which seven are found in Australia and its surrounding coastal waters.

FAMILY TYPHLOPIDAE (PAGES 11–13)
Characterized by shiny worm-like bodies with a blunt head and a short tail terminating in a spine. The eyes are reduced to dark spots and the scales are uniform around the body: there are no enlarged scales on the belly. The mouth is small and situated beneath an overhanging snout. When handled, these small snakes have a tendency to emit a strong, pungent odour. There are 30 currently recognized species in Australia, all of which apparently eat ant and termite pupae, larvae and eggs. They are fossorial and are usually found beneath ground debris, rocks, or in ant nests and termitaria.

FAMILY BOIDAE (PAGES 14–21)
With the exception of the two *Aspidites* species, all Australian pythons have heat-sensing pits in some of the scales of the lips. They all have 30 or more scale rows around the middle of the body, and the belly scales are larger than the other body scales. All pythons possess spurs (which are the vestiges of ancestral hindlimbs) on either side of the vent. Females incubate and protect their eggs by curling around them until hatching occurs. They are muscular, usually heavily built, slow moving and non-venomous.

FAMILY ACROCHORDIDAE (PAGES 22–23)
The three species of file snakes in this family are totally aquatic and have large robust bodies with a 'baggy' skin. Unlike the paddle-like tail of the typical sea-snakes, the tail of the file snakes is narrow and prehensile to permit its use as an entwining 'anchor'. The scales are small and very rough and rasp-like. There are 80 or more scale rows around the body and the belly scales are no larger than the scales over the back. The nostrils are valvular and are situated high on the snout. All species bear live young.

FAMILY COLUBRIDAE (PAGES 24–29)
Found across northern Australia and down the east coast, these snakes either have solid teeth throughout and are non-venomous, or have solid teeth at the front of the mouth and enlarged, grooved, poison-conducting fangs at the back of the mouth. All species have fewer than 30 scale rows around the mid-body and the belly scales are larger than the other body scales, being as wide as the belly itself. There are one or two loreal scales present, except in the White-bellied Mangrove Snake. The tail is generally cylindrical.

Family Elapidae (pages 30–63)
All elapid snakes are venomous although most are only mildly so and not considered to be dangerous. They all have fixed hollow fangs in the front of the upper jaw, connected on each side by a duct to the venom gland. No elapids have loreal scales and there are 15 to 23 scale rows around mid-body, excluding the single row of broad belly scales. Some species give birth to live young; the others lay eggs.

Family Hydrophiidae (pages 64–69)
The reptiles of this family are completely aquatic sea-snakes with vertically compressed, paddle-shaped tails and nostrils that are situated on the top of the snout with valvular flaps which close when the snake is submerged. There are no enlarged belly scales in snakes of this family, the normal body scales continuing right round under the belly. All species give birth to their young underwater.

Family Laticaudidae (page 70)
While essentially marine snakes, these 'sea kraits' do come on to land, particularly to lay their eggs. All are characterized by numerous black cross-bands and laterally placed nostrils. The belly scales are larger than the other body scales.

SUBORDER SAURIA
Lizards are represented by some 17 families worldwide, of which five are found in Australia.

Family Gekkonidae (pages 80–89)
The majority of geckos are essentially nocturnal and all are capable of vocalization. They have well-developed fingers and toes, and a tail that is easily cast off in emergencies. The eyes are prominent. The scales are small, granular, non-shiny and do not overlap. In many species the digits end in an expanded pad that aids climbing on vertical and overhanging surfaces. All Australian species lay eggs.

Family Pygopodidae (pages 90–95)
These lizards are snake-like or worm-like with no trace of forelimbs. The hindlimbs are represented by, at most, scaly flaps which are situated at either side of the vent. Most species have external ear openings and all have a broad fleshy tongue with a notch at the tip. The tail is long and easily cast off. All species lay eggs and can vocalize.

Family Varanidae (pages 96–102)
These lizards are characterized by a dull, rough, apparently loose-fitting skin, a long flat head and a long neck. The tongue is long and deeply forked, and constantly flicked in and out. The tail is long and slender and the limbs well developed with sharp claws. All species lay eggs.

Family Agamidae (pages 103–112)
The skin of these 'dragon-lizards' usually has a rough, dull appearance, often with tubercles, prickles or spines. The tail is long and tapered and there are four well-developed limbs. The tongue is broad and fleshy. All species are diurnal and lay eggs.

FAMILY SCINCIDAE (PAGES 113–139)
With well over 300 species in Australia, it is difficult to generalize on the features of this family. Ranging in size from a few centimetres to over 60 cm they come in many forms. Most have smooth shiny scales and a long tail that can be easily cast off. Others have a rough prickly skin. Although four limbs are usual there are species with only two limbs, while still others are completely limbless. Some species lay eggs and the others bear live young.

Field hints

Reptiles are protected in all States of Australia and it is illegal to disturb them in any way; however this attitude of 'protecting' individual animals does very little for the conservation of species. Land-clearing and road-kills in a single year destroy far more reptiles than all the 'disturbing' or collecting by reptile enthusiasts. Nevertheless, the reader should be aware that many of the conservation authorities find it far easier to harass and prosecute individuals than property developers, other large organizations or Government bodies. Any interaction with reptiles should be approached with this in mind. Under no circumstances retain reptiles in your possession unless you are acting within the regulations of the State and/or you have the required permits.

Permission should always be obtained when going on to private land. Country folk are usually very friendly but farmers understandably take exception to strangers wandering about their properties without asking first.

Wear sensible and suitable clothing and footwear. Keep your eyes and ears open, and move slowly and quietly. Study the area ahead, looking for basking or foraging animals. In rocky areas examine crevices and ledges (but do not destroy crevice habitats). In short, be observant! Binoculars can be of use in some situations where the terrain is reasonably open. When you see or hear a reptile, stop. If you are fortunate it may not have seen you or you may still be far enough away for it not to race for cover. Only experience will tell you how close you can get to a particular species before it flees.

Larger reptiles can often be identified by observation but most smaller species will require close scrutiny or even capture to identify positively. A notebook should be carried as it will often be easier to jot down distinguishing features on the spot and refer to this guide later.

Where to look

Reptiles may be found in almost all parts of Australia, from inner suburbs of the major cities to Mount Kosciusko. But, unlike birds which are highly visible, numerous and usually very vocal, reptiles tend to be more secretive and a high proportion are fossorial or nocturnal species. Nevertheless it can be very rewarding to seek them out and observe them in their habitats.

All environments from rainforest to desert support a reptile fauna, but finding a particular species depends very much upon the weather conditions, the time of year and the time of day. Except in the tropical north, most reptiles become dormant during the winter months. Spring is one of the best times to see reptiles, when many species are actively seeking mates and foraging for food after winter dormancy. While all reptiles seek warmth, excessive heat kills them very quickly, hence you are unlikely to see animals out in the middle of the day during hot weather. Most will be active in the early morning or late afternoon. However during mild or overcast weather they can be active all day.

When inactive, some reptiles shelter beneath rocks, logs, sheets of tin, timber, rubbish, leaf litter, loose bark and many other materials that afford them some protection. Others shelter in burrows, logs or under the surface layers of soil.

You are more likely to see lizards than snakes, since many are diurnally active and more visible. They are also more numerous than snakes. Many fossorial species are not seen in the open at all, and these must be searched for by looking beneath close-fitting objects on the ground. Always carefully replace any rocks or other material lifted and keep any disturbance to a minimum.

Road-hunting by day or night can often result in spotting reptiles either basking on or near the road, or moving across the road itself. Choose roads which have little traffic and that pass through likely habitat. By driving slowly along such roads in the right conditions, a surprising variety of reptiles can often be seen.

Reptiles enjoy more tolerance now than has previously been the case. Perhaps Kermit and the ninja turtles have played a part but one would hope that the greater awareness is due to television documentaries, the availability of books on the subject, and the education of children in schools. People may still not like them but at least there is a greater reluctance in some regions to kill them.

In the author's view people should be able to keep reptiles if they so wish and if they are capable of doing so. Much of Australian wildlife legislation is more concerned with putting restrictions on the keeping of reptiles than with the actual conservation of them.

How to use this book

This book includes 223 of the more than 760 reptiles known from Australia. These have been selected to provide a representation of the families, and to include those species most likely to be encountered or which are larger and more obvious.

Common names have been included for all species and most of these follow Ehmann (1992). Scientific names in the main follow Cogger, *Reptiles and Amphibians of Australia* (1992).

All measurements in this book are given in centimetres (cm). The measurements at the head of each species account represent the recorded range of lengths of sexually mature adults (snout to tail-tip).

Each snake species is cited as 'non-venomous' (harmless) or 'venomous'. Those shown as venomous are recorded in one of three categories:

Venomous but not regarded as dangerous. A bite would produce either no symptoms or only mild local symptoms, with no danger to health.

Venomous. A bite may require treatment. In some cases a bite from such a snake may require attention. A person bitten by one of these species should be given appropriate first aid and monitored until symptoms have disappeared.

Venomous and dangerous. Capable of inflicting a potentially fatal bite. A bite from one of these species must be regarded as life-threatening. In all such cases, first aid must be administered immediately and the victim transported to hospital.

Identification in the field

Note the main features of the animal you wish to identify: the general profile, the number of limbs, and their degree of development. What is the tail length compared with that of the body? Are the scales rough or smooth? Does it have uniform colouration? If not, are there blotches, bands or stripes? Try to identify the animal as a snake or lizard. This is usually obvious but you need to remember that many of the pygopodids and some skinks are very snake-like. Snakes, however, do not have ear openings; lizards do.

You will note that in the snake family descriptions in the section on classification (page 5), reference is made to the number of scale rows around the body. Clearly it is possible to count these rows only when the snake is dead or immobilized, and it should be noted that the count should be made at or near the middle of the body because the number may vary towards the head or tail. The (usually) single row of wide belly scales is excluded from this count.

If you are able to identify the animal to the family level, i.e. as a dragon, gecko, etc., turn to the appropriate section of the book. If you cannot identify the animal to the family level read the family descriptions on pages 5–8. Look through the illustrations until you find one that most closely resembles the animal you want to identify. Read the description, note the size and look at the map. If you still cannot identify your animal, don't despair. This book does not include all the reptiles of Australia. You may have a species not included here. It may be that the animal has an unusual colour pattern. Colour is often unreliable in identification, since there can be very great variation within a species, between sexes, and between juveniles and adults of the same species. Even experts have problems occasionally. A list of other books is given on page 141 should you wish to take the identification process further.

Snake bite: dos and don'ts

It is worth remembering that snakes do not go out of their way to bite people. Snake-bite incidents are usually the result of an accidental encounter – a clumsy approach or careless handling on the part of a would-be snake-catcher.

Remember:
- not all snakes are venomous • not all venomous snakes inflict a potentially fatal bite • venom may not have been injected

Do not:
- cut, suck or wash the site of the bite • apply a tourniquet • give the victim alcohol, coffee or other stimulants • apply Condy's crystals (potassium permanganate) or ice to the bite (All of the old 'remedies' above have been totally discredited.)

Try to keep the victim calm and at rest, and transport him or her quickly to the nearest medical centre. If access to the victim by vehicle is difficult, use a stretcher if possible. Failing this the victim should walk but not run. If the snake has been killed during the biting incident, take it to the medical centre for identification, but remember it is not worth risking a further bite in order to secure a snake corpse for this purpose.

All hikers and bush-walkers should familiarize themselves with the use of a pressure bandage as a first-aid measure in the treatment of snake bites on a limb.

BLIND SNAKES Family TYPHLOPIDAE

These snakes are characterized by shiny worm-like bodies with a blunt head and a short tail terminating in a spine. The eyes are reduced to dark spots and the scales are uniform around the body, that is, there are no enlarged scales on the belly. The mouth is small and is situated beneath an overhanging snout. When handled they have a tendency to emit a strong, pungent odour. There are 30 currently recognized species in Australia, all of which apparently eat ant and termite pupae, larvae and eggs. All members of the family are fossorial and are usually found beneath ground debris, rocks or in ant nests and termitaria.

Southern Blind Snake *(Ramphotyphlops australis)* 25-50 cm

Dick Whitford

This blind snake is of moderate to robust build with 22 scales around the body. It is purplish-grey to purplish-brown above and white underneath. Where the dorsal colour meets the white belly colour there is a ragged merging. The snout is rounded when viewed from above or in  profile and the rostral scale is slightly longer than it is broad. It occurs in southern Western Australia, South Australia, the southern part of the Northern Territory, northern Victoria and south-western New South Wales. Females lay from two to 11 eggs in midsummer. HARMLESS.

Prong-snouted Blind Snake *(Ramphotyphlops bituberculatus)* 30-45 cm

The head is distinctly trilobed from above and pointed in profile in this slender to moderately built snake. It has 20 scales around the body and the colour ranges from brown to almost black with a white to pinkish belly. The rostral scale is longer than broad. It occurs in southeastern Western Australia, South Australia, northern Victoria, New South Wales except the east coast, southern Queensland and southeastern Northern Territory. From two to nine eggs are laid in summer. HARMLESS.

Northern Blind Snake *(Ramphotyphlops diversus)* 25-35 cm

This snake is of slender to moderate build with 20 scales around the body. The rostral scale is longer than broad and the snout is rounded from above and when seen in profile. It is brown to blackish-brown above, grading quite smoothly into the cream undersurface. It occurs in the northern regions of Western Australia, the Northern Territory and Queensland. HARMLESS.

Blackish Blind Snake *(Ramphotyphlops nigrescens)* 35-75 cm

This usually has a black patch on either side of the vent. It is of moderate to robust build with 22 scales around the body. The colour ranges from pink-brown to almost black, with the base of each scale usually paler, forming a vague reticulation. The rostral scale is slightly longer than broad and the snout is rounded from above and in profile. It occurs in south-eastern Queensland, eastern New South Wales and eastern and southern Victoria. Females lay from five to 20 eggs in midsummer. HARMLESS.

Proximus Blind Snake *(Ramphotyphlops proximus)* 50-75 cm

Of robust build, this snake has 20 scales around the body and a rostral scale that is almost circular in shape. The snout is bluntly trilobed when viewed from above and angular in profile. It is dark brown in colour and sometimes has a dark patch on either side of the vent. It occurs in northern Victoria, the interior of New South Wales and eastern Queensland. Females lay a clutch averaging 13 eggs during the summer. HARMLESS.

PYTHONS Family BOIDAE

Pythons are non-venomous. Except for *Aspidites* they have heat-sensing pits in some of the scales of the lips which help them to locate warm-blooded prey at night. All pythons possess spurs on either side of the vent which are the vestiges of ancestral hindlimbs, and they all have 30 or more scale rows around the middle of the body. The females incubate and protect their eggs by coiling around them until hatching occurs.

Black-headed Python *(Aspidites melanocephalus)* 150-250 cm

The distinctive black head and neck of A. melanocephalus.

This is easily recognizable by the glossy black head and neck. The black pigmentation ends abruptly at the neck and is replaced by a body colour of cream to reddish-brown with dark-brown cross-bands. The head scales are large and symmetrical. It lacks any heat-sensitive pits on the lips. Found in the northern parts of Western Australia, the Northern Territory and Queensland, it inhabits woodlands, shrublands and grasslands. It is terrestrial, sheltering within crevices, caves, hollow logs or burrows. It is mainly nocturnal but is sometimes active during the day, feeding mainly on reptiles but also eating small mammals and birds. From six to 12 eggs are laid in October and November. NON-VENOMOUS.

Woma *(Aspidites ramsayi)* 150-300 cm

The head of this snake has a somewhat pointed appearance and, like the Black-headed Python, it lacks heat-sensing pits on the lips. It is of robust build, olive to red-brown in colour with numerous darker bands across the body. In some areas there is a black patch over the eyes.

The head scales are large and symmetrical. It occurs through the arid areas of all states except Victoria and favours sandy areas where it shelters in burrows. Nocturnal and terrestrial, it feeds mainly on reptiles and mammals but sometimes takes birds. From four to eight eggs are laid in late spring or early summer. NON-VENOMOUS.

Green Python *(Morelia viridis)* 120-200 cm

Juvenile with yellow colouration

With a colouration of dull to bright emerald-green it would be difficult to confuse this snake with any other. It has a large head which is distinct from the neck and there is a white or bluish vertebral marking from the nape to the tail. The head scales are small and irregular.

Juveniles can be yellow, orange, brick-red or brown but eventually change to green. It is nocturnal and arboreal and found in rainforest, monsoon forest and bamboo thickets on northern Cape York Peninsula. It eats reptiles and mammals. From 11 to 26 eggs are laid. (This python is sometimes referred to as *Chondropython viridis.*) NON-VENOMOUS.

Olive Python *(Liasis olivaceus)* 250-650 cm

The colour of this nocturnal python varies from pale fawn to dark olive-brown above with a creamy white belly. The head scales are large and symmetrical. It prefers rocky habitats and is found in northern Australia from about Broome in Western Australia to western Queensland. An isolated population also occurs in the Pilbara region of Western Australia. It feeds on mammals, birds and reptiles. From 11 to 21 eggs are laid in October and November. NON-VENOMOUS.

Water Python *(Liasis fuscus)* 150-300 cm

Head of Water Python.

Very iridescent dark-brown above with whitish lips and throat, this python is yellowish underneath except for the tail, which is dark brown. The head scales are large and symmetrical. It occurs through northern Australia from Western Australia to Queensland and is generally found near water. The main food items are mammals and birds but other reptiles are also taken. From six to 23 eggs are laid in October and November. NON-VENOMOUS.

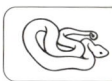

Eastern Small-blotched Python *(Antaresia maculosa)* 75-120 cm

This solidly built python is a light brown to medium brown above, lighter on the sides, with numerous dark, ragged-edged blotches along the body: these may sometimes merge to form zigzag stripes. The head scales are large and symmetrical. Found in eastern Queensland and north-eastern New South Wales, it occurs in woodlands and sclerophyll forests, favouring rocky outcrops. It feeds mainly on small mammals and reptiles. From four to 20 eggs are laid in October or November. NON-VENOMOUS.

Children's Python *(Antaresia childreni)* 75-100 cm

This python is relatively slender in appearance and ranges in colour from pale brown to a dark purplish-brown which is lighter on the lower sides. If a pattern is present it consists of a series of smooth-edged darker blotches which are not very prominent. The head scales are large and symmetrical. Found in northern Australia from north-western Queensland to the Kimberley region of Western Australia, it occurs in a variety of habitats including rock outcrops and woodlands. It feeds on small mammals, reptiles and frogs. From four to nine eggs are laid. (This python is sometimes referred to as *Liasis childreni*.) NON-VENOMOUS.

Pygmy Python *(Antaresia perthensis)* 40-60 cm

Certainly the smallest python found in Australia and probably in the world, this is found in crevices, burrows and large termite-mounds in the Pilbara region of Western Australia. It has a prominent head, distinct from the neck, and a body colour from light brown to a reddish-brown, with darker obscure variegations. The head scales are large and symmetrical. It feeds on reptiles and small mammals. From two to five eggs are laid in November or December. (This python is sometimes referred to as *Liasis perthensis*.) NON-VENOMOUS.

Large-blotched Python *(Antaresia stimsoni)* 75-100 cm

The smooth-edged dark-brown blotches which sometimes merge to give a banded appearance are a feature of this python which is light brown to yellowish-brown. The head scales are large and symmetrical. Often associated with stony ranges or outcrops, it is widely distributed through the interior of Australia from Western Australia to western New South Wales and Queensland. It feeds predominantly on mammals and reptiles. From six to 19 eggs are laid between August and November. (This python is sometimes referred to as *Liasis stimsoni*.) NON-VENOMOUS.

Amethystine Python *(Morelia amethistina)* 350-850 cm

Australia's largest snake is named in reference to the milky iridescent sheen of the scales. It has a slim elongate body and the head is distinct from the neck. The body colour ranges from fawn to olive-brown and there are numerous darker-brown or black variegated markings.

The belly is whitish. The head scales are large and symmetrical. It occurs in north-eastern Queensland and on the islands of Torres Strait, where it is found in rainforest, open forest and monsoon forest. It is arboreal and feeds on mammals as large as wallabies. Eggs are laid in December and clutch sizes range from seven to 12. NON-VENOMOUS.

Centralian Carpet Python *(Morelia bredli)* 150-260 cm

This species has a large head which is distinct from the neck, and a robust body which is reddish to dark-brown in colour with numerous irregular pale fawn, dark-edged blotches or bars. The belly is white with brown or grey variegations and there is a pale streak running from the eye to the base of the head. The head scales are small and irregular. It is found in the arid southern parts of the Northern Territory and inhabits rocky ranges and large trees in watercourses. It feeds on mammals and birds. The clutch size is 13 to 47 eggs. NON-VENOMOUS.

Oenpelli Rock Python *(Morelia oenpelliensis)* 200-450 cm

Slender and with a long prehensile tail, this python has a long head which is distinct from the neck. The body is a fawn colour becoming lighter on the sides. There are numerous darker blotches or streaks that are roughly aligned in four or five rows along the body. A dark-brown streak runs from the snout or eye to the corner of the jaw. Most of the head scales are large and symmetrical. Restricted to the Arnhem Land escarpment and outliers in the Northern Territory, it inhabits cliffs and gorges. It is nocturnal and feeds on mammals and birds. From two to ten eggs are laid in October and November. NON-VENOMOUS.

Carpet Python *(Morelia spilota variegata)* 150-400 cm

This python is extremely variable in colouration and it may be that more than one species is involved. Four distinct forms are recognized. A form from the Northern Territory and Kimberley region of Western Australia is blackish-brown, paling to red-brown on the sides, and has dark-edged, ragged yellow to cream bands across the body. The Murray/Darling River form is reddish-brown with a series of grey blotches that are dark-edged and often more or less paired dorsally. The rainforest form, which inhabits the rainforest of north-eastern Queensland, is black with distinctive yellow blotches and stripes running along the body. The eastern form, which is the most widespread, is pale to dark brown with numerous irregular dark-edged cream blotches. Although mainly nocturnal, the Carpet Python is often found basking during the day. It occurs from the semi-arid inland regions to rainforest areas and feeds mainly on mammals but will also eat reptiles and birds. From nine to 54 eggs are laid between November and January. This species has a well-deserved reputation for being 'bad-tempered'; it usually hisses loudly if disturbed and does not hesitate to bite. NON-VENOMOUS.

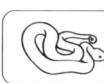

Diamond Python *(Morelia spilota spilota)* 150-400 cm

The very distinctive colouration makes this python easily recognizable. It is olive-black above, usually with a cream or yellow spot on most scales. Some of the spots are arranged in the form of rosettes along the length of the body. The head scales are small and irregular. It occurs along the New South Wales coastal strip from the Victorian border to about Coffs Harbour. Found in rainforest and heavily timbered areas, it is also common in some suburban areas. Nocturnal and usually arboreal, it feeds mainly on mammals and other reptiles although birds are also eaten. Eggs are laid in December and January and clutch sizes range from nine to 54. NON-VENOMOUS.

West Australian Carpet Python *(Morelia spilota imbricata)* 150-250 cm

This python is distinguished by the strongly overlapping and quite elongate dorsal scales. The body is greenish-brown to black-brown with black-edged light blotches that are elongated and interconnected. The head scales are small and irregular. It is found in the south-western region of Western Australia and feeds on mammals, reptiles and birds. NON-VENOMOUS.

FILE SNAKES Family ACROCHORDIDAE

These snakes are totally aquatic and have large robust bodies with a 'baggy' skin. Unlike the paddle-like tail of the typical sea-snakes, the tail of the file snakes is narrow and prehensile to permit its use as an entwining 'anchor'. There are 80 or more scale rows around the body and the belly scales are no larger than the scales over the back. The scales are small and very rough and sharp-pointed. The nostrils are valvular and situated high on the snout. All species bear live young. File snakes are not venomous, and restrain their slippery fish prey by biting them and gripping them with the rasp-like skin of their coils.

Arafura File Snake *(Acrochordus arafurae)* 150-250 cm

As in other file snakes, the head is barely distinct from the neck and the loose flaccid body skin has small spinose scales. The body is dark brown with numerous pale blotches that may merge towards the head. It is most common in freshwater rivers and billabongs but may also be encountered in the ocean or in estuaries. Nocturnal in habit, it shelters among aquatic vegetation and underneath overhanging banks; it is a fish-eater. It occurs in northern Australia from Western Australia to Queensland. From 11 to 27 young are born in February. NON-VENOMOUS.

Little File Snake *(Acrochordus granulatus)* 50-120 cm

This rather slender file snake has small spinose scales and a strongly compressed tail. It is brownish-black above with numerous narrow pale fawn cross-bands. It is a marine species and is found across northern Australia from Western Australia to Queensland. It occurs in coastal waters and favours areas such as mangrove estuaries, reef flats and mudflats. It is nocturnal and feeds on small fish. From two to 12 young are produced. NON-VENOMOUS.

COLUBRID SNAKES Family COLUBRIDAE

These snakes have solid teeth and are either non-venomous with solid teeth throughout, or slightly venomous with solid teeth at the front of the mouth and enlarged, grooved, poison-conducting fangs at the rear of the mouth. With the exception of the White-bellied Mangrove Snake all our colubrids have one or two loreal scales - elapid snakes lack loreals - and all species have fewer than 30 scale rows at mid-body.

Northern Brown Tree-snake *(Boiga fusca)* 100-200 cm

G.E. Schmida

This species has a very broad head, quite distinct from the narrow neck, and large eyes with vertical pupils. It is identified by its cream to pale reddish-brown colour with prominent broad, irregular, dark red-brown bands. It occurs in northern Australia from the Kimberley region of Western Australia to the north-west coast of Queensland where it is found in woodlands, vine thickets and monsoon forests, seeming to prefer rocky areas. It is nocturnal and feeds on other reptiles, birds and small mammals. A clutch of four to 12 eggs is laid each year. VENOMOUS BUT NOT REGARDED AS DANGEROUS.

Eastern Brown Tree-snake *(Boiga irregularis)* 100-200 cm

G.A. Hoye

Like the Northern Brown Tree-snake, this species has a very broad head and large prominent eyes with a vertical pupil. It is red-brown to brown above with numerous narrow dark cross-bands, and salmon-coloured on the belly. It occurs along the east coast of Australia from  Cape York Peninsula to the central coast of New South Wales. Nocturnal and arboreal, it is found in wet and dry sclerophyll forests, rainforests, mangroves and heaths. When disturbed it becomes very aggressive, raising the forepart of its body into a series of S-shaped curves and striking repeatedly. It feeds on birds, lizards, small mammals and occasionally frogs. From four to 12 eggs (clutches averaging six) are laid each year. VENOMOUS BUT NOT REGARDED AS DANGEROUS.

Green Tree-snake (*Dendrelaphis punctulata*) 100-200 cm

Blue form (see additional photographs on page 27)

This slender snake has a whip-like tail and large eyes with round pupils. It shows considerable geographic colour variation. The form found in New South Wales and most of Queensland is grey to olive-green in colour and usually yellow underneath, particularly on the throat and neck. In northern Queensland some individuals may be dark-brown to black, or blue. In the Northern Territory and Western Australia it has a golden-yellow colour with a bluish head. It is diurnal and found in a wide range of habitats along the coastal regions of northern and eastern Australia from northern Western Australia to south-eastern New South Wales. In some areas it extends inland along the rivers. When disturbed, it can release a strong-smelling odour from the vent and may inflate the body and neck threateningly, exposing the blue skin between the scales. Although feeding mainly on frogs, it will also eat reptile eggs and small mammals. The annual clutch is five to 14 eggs. NON-VENOMOUS.

Yellow form

Green form

Slate-grey Snake *(Stegonotus cucullatus)* 100-150 cm

This snake has smooth, polished scales and is dark grey to black above, yellow to white below. The upper lip is usually paler than the rest of the head. It occurs in northern Queensland and northern areas of the Northern Territory. A nocturnal species usually associated with streams and lagoons, it is aggressive when disturbed and will bite repeatedly. It may also release a strong odour from its anal glands. It feeds mainly on other reptiles and frogs but small mammals are sometimes taken. Reptile eggs are also eaten. The annual clutch is five to 16 eggs. NON-VENOMOUS.

Keelback *(Tropidonophis mairii)* 50-100 cm

One of the colour variations.

The scales of this snake are strongly keeled, hence the name, and it has the unusual capacity (for a snake) of being able to discard its tail if it is grasped. It has large eyes and a loreal scale (which distinguishes it from the dangerous elapid, the Rough-scaled Snake *Tropidechis carinatus*). It ranges in colour from grey, brown, olive and reddish to black, with numerous dark-tipped scales which may form irregular cross-bands. Underneath it may be cream, brown, salmon or olive-green. It is a semi-aquatic snake usually found near streams, lagoons or swamps in the coastal regions from northern Western Australia to northern New South Wales. It feeds almost exclusively on frogs and is apparently able to eat young cane toads without ill effect. It has also been recorded to take fishes and reptile eggs. The Keelback has a very short activity period, only around dusk. It can emit an unpleasant odour from its anal glands if disturbed. The annual clutch is five to 15 eggs. NON-VENOMOUS.

Macleay's Water Snake *(Enhydris polylepis)* 60-100 cm

This species is iridescent dark brown above and along the flanks, and yellow or cream speckled with black on the lower sides. The belly is cream or yellow with a black stripe on the undersurface of the tail. It has small eyes set towards the top of the head, and valvular nostrils.

Found in north-eastern Queensland and the northern parts of the Northern Territory it is a nocturnal, aquatic species that inhabits creeks, swamps and rivers where it feeds on fish and frogs. It gives birth to ten to 15 young in mid-January. VENOMOUS BUT NOT REGARDED AS DANGEROUS.

White-bellied Mangrove Snake *(Fordonia leucobalia)* 66-100 cm

This is the only Australian colubrid lacking loreal scales. It is variable in colour and colour phases range from black, brown or yellow to red, with scattered white, yellow, grey or pink scales. It has a blunt head, its eyes are small and it has valvular nostrils as an adaptation to its semi-aquatic life-style. It is nocturnal and distributed through northern Australia from Cape York Peninsula in Queensland to northern Western Australia, where it is found on exposed mudbanks at low tide, or in mangrove swamps. It feeds on crabs, and if a crab is too large to eat whole it will twist off and eat the legs one by one before eating the body. It gives birth to three to 13 young in March or April. VENOMOUS BUT NOT REGARDED AS DANGEROUS.

Two of the colour phases.

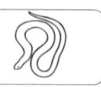

ELAPID SNAKES Family ELAPIDAE

The elapid snakes are all venomous although most are only mildly so and not considered to be dangerous. They all have fixed hollow fangs in the front of the upper jaw and these are connected by a duct to the venom gland. They have between 15 and 23 scale rows at mid-body (excluding the single row of broad ventral scales down the belly) and none have loreal scales. Some species lay eggs and some give birth to live young.

Southern Death Adder *(Acanthophis antarcticus)* 40-105 cm

The distinctive Death Adder head.

The broad triangular head, narrow neck and stout body, and thin tail ending in a spine make this snake readily identifiable. Its colour ranges from grey to brownish-red, usually with distinct cross-bands. The white or cream edges of the labial scales form prominent bars on the lips. Occurring over most of Queensland and New South Wales, as well as southern Western Australia and South Australia and the northern areas of the Northern Territory, it inhabits woodlands and heathlands through to wet sclerophyll rainforest. It is sedentary and burrows into leaf litter or soft soil, feeding on lizards, small mammals and birds which are often caught by positioning the tail close to the head and twitching its tip. When the prey animal comes in to catch what appears to be an insect or grub, the snake strikes with great speed and accuracy. The female gives birth to two to 30 young. VENOMOUS & DANGEROUS. CAPABLE OF INFLICTING A POTENTIALLY FATAL BITE.

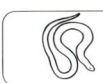

Northern Death Adder *(Acanthophis praelongus)* 40-70 cm

This snake has a broad triangular head, narrow neck and stout body with a thin tail terminating in a soft spine: the scales are keeled. The colour varies from grey to brownish-red, usually with distinct cross-bands. It has a lighter build than the Southern Death Adder and its head scales  are more rugose. Found in northern Australia from the Kimberley region of Western Australia across to Queensland, it usually inhabits grasslands, woodlands or rocky outcrops. It feeds mainly on lizards. VENOMOUS & DANGEROUS. CAPABLE OF INFLICTING A POTENTIALLY FATAL BITE.

Desert Death Adder *(Acanthophis pyrrhus)* 50-75 cm

The scales of this snake are very strongly keeled. Although it has a similar broad triangular head, narrow neck and robust body with a thin tail terminating in a soft spine, it is more slender than the other two death adders. The colour varies from light reddish-brown to bright  orange-brown with lighter cross-bands. Occurring in the arid regions of western and central Australia, it inhabits spinifex-dominated regions in sandy areas or rocky outcrops. It lies concealed in loose red sand and feeds on lizards and small mammals. The females give birth to up to 13 young which are born in summer. VENOMOUS & DANGEROUS. CAPABLE OF INFLICTING A POTENTIALLY FATAL BITE.

Pygmy Copperhead *(Austrelaps labialis)* 50-70 cm

This copperhead is dark grey to black with the lips and side of the head prominently barred with cream. There may be a narrow dark vertebral stripe on some specimens and there are two distinct pale parallel lines running along the lower sides. Restricted to South Australia, it is found on Kangaroo Island, the Fleurieu Peninsula and the western and southern areas of the Mount Lofty Ranges, where it inhabits woodlands, heath shrublands and dry sclerophyll forest. It feeds mainly on lizards but is known also to take frogs and small mammals. An average of seven live young are produced in a litter. VENOMOUS & DANGEROUS. CAPABLE OF INFLICTING A POTENTIALLY FATAL BITE.

Highland Copperhead *(Austrelaps ramsayi)* 80-115 cm

This copperhead has conspicuous white edging to the scales of the upper lips. The snout is paler than the rest of the head and the body varies in colour from greyish-brown to black. The scales on the lower sides are white to cream. Found in the cool and cold parts of the Great Dividing Range in New South Wales and eastern Victoria, it is usually associated with streams, swamps and seepages, especially where there is a dense ground cover of tussock grass. It feeds on lizards and frogs. Females produce from nine to 30 young. VENOMOUS & DANGEROUS. CAPABLE OF INFLICTING A POTENTIALLY FATAL BITE.

Lowland Copperhead *(Austrelaps superbus)* 100-170 cm

This copperhead ranges in colour from brown, reddish-brown, and grey to almost black. The scales on the lower sides may be cream, yellow, pink or orange and some specimens have a narrow dark vertebral stripe. The lips and side of the head are marked with dark and pale bars but not as prominently as in the Highland Copperhead. It occurs in Tasmania, southern Victoria and south-eastern South Australia, and also on some Bass Strait islands. It feeds on lizards and frogs but has also been known to take small mammals. Females produce from nine to 40 young. VENOMOUS & DANGEROUS. CAPABLE OF INFLICTING A POTENTIALLY FATAL BITE.

White-crowned Snake *(Cacophis harriettae)* 25-50 cm

This species is distinguished by a white to cream stripe around the head, widest on the nape where it forms a distinct collar that is at least four scales wide. The upperparts are grey to dark brown and the belly is dark grey. Found in coastal areas of south-eastern Queensland and north-eastern New South Wales, usually in wet sclerophyll forest and rainforest areas, it is a nocturnal species which feeds on small lizards and reptile eggs. It has an unusual defensive posture in which it raises the forebody off the ground and points the head downwards to display the top-of-the-head markings. The annual clutch is between two and ten eggs. VENOMOUS BUT NOT REGARDED AS DANGEROUS.

Dwarf Crowned Snake *(Cacophis krefftii)* 20-35 cm

In this relative of the White-crowned Snake, the white to cream stripe around the head and nape is only one to two scales wide on the nape. It is black or brownish-black above and white underneath, with a distinctive black saw-tooth pattern laterally. Occurring from the central coastal region of New South Wales to south-eastern Queensland, it is found in wet sclerophyll forest and rainforest. It is nocturnal and feeds on small lizards. Females lay clutches of from two to five eggs in November or December. VENOMOUS BUT NOT REGARDED AS DANGEROUS.

Golden-crowned Snake *(Cacophis squamulosus)* 40-70 cm

In this crowned snake, the yellowish stripe (the 'golden crown') around the head extends well back on to the nape but does not meet in the mid-line. The body is dark brown to blackish above and pink to orange below, with black spots. The head is distinct from the neck and relatively large. Found in wet sclerophyll forest and rainforest from the central coast of New South Wales to the central coastal areas of Queensland, it is a nocturnal species living under well-embedded rocks. It feeds on lizards and reptile eggs. The annual clutch is two to 15 eggs. VENOMOUS BUT NOT REGARDED AS DANGEROUS.

Black Whipsnake *(Demansia atra)* 100-180 cm

As in all whipsnakes, this species has a slender body and long tapering tail. It is light or dark olive-brown to blackish above, and each scale is edged with darker colour. There is often a reddish-brown tinge on the posterior parts of the body. The head is deep and narrow and distinct from the neck. Active and diurnal, this large species is found in northern Australia from the Kimberley region of Western Australia to central eastern Queensland. It inhabits dry open forests or woodlands where there is a grassy understorey. It feeds mainly on lizards but also takes frogs. The annual clutch is four to 20 eggs.
VENOMOUS: A BITE MAY REQUIRE TREATMENT.

Olive Whipsnake *(Demansia olivacea)* 65-85 cm

Slightly less slender than other whipsnakes, this species has a head barely distinct from the neck. The top and sides of the head are usually marked with dark blotches. The body is olive-grey, brownish or reddish, often with a dark spot at the base of each scale. Occurring in northern Western Australia, the northern part of the Northern Territory and north-western Queensland, it is a diurnal fast-moving snake that feeds on lizards and reptile eggs. Between three and four eggs constitute the normal clutch. VENOMOUS BUT NOT REGARDED AS DANGEROUS.

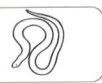

Collared Whipsnake *(Demansia torquata)* 50-85 cm

G.E. Schmida

Most slender and whip-like of the whipsnakes, the Collared Whipsnake is olive-brown to grey, often with a reddish tinge. The head is darker, with one or more black bands edged with yellow. There is a narrow dark line across the front of the snout and a pale-edged dark 'comma' around the eye. Occurring in north-eastern Australia from the New South Wales border through Queensland and the Northern Territory to north-eastern Western Australia, in grassy woodlands, heathlands and hummock grasslands, this is a fast-moving snake that feeds on lizards. Females lay clutches of between two and eight eggs in spring. VENOMOUS BUT NOT REGARDED AS DANGEROUS.

Yellow-faced Whipsnake *(Demansia psammophis)* 80-100 cm

R.W.G. Jenkins

This whipsnake has a distinctive dark 'comma' with a creamy-yellow margin around the eye, and a pale-edged dark band across the snout. The body colour in this slender and very attractive species ranges from grey to olive-brown, often with a reddish-brown flush on the nape and foreparts of the body. The two subspecies found in Western Australia have dark-edged scales giving them a very reticulated appearance. Widespread over continental Australia except in the tropical north, it feeds mainly on lizards but it is known also to eat frogs as well as reptile eggs. Females lay three to nine eggs in a clutch and communal egg-laying has been recorded. VENOMOUS BUT NOT REGARDED AS DANGEROUS.

De Vis's Banded Snake *(Denisonia devisi)* 40-60 cm

Sometimes mistaken for a small death adder, this snake has a robust body and a broad depressed head which is distinct from the neck. The body is light brown above with dark irregular cross-bands on the body and tail. The head is dark brown with conspicuously barred lips.

Found in central western New South Wales through to central northern Queensland, it occurs in low-lying moist areas in woodlands and shrublands. It is nocturnal and feeds mainly on frogs, although lizards are also taken. Litters are normally of three to nine young. VENOMOUS A BITE MAY REQUIRE TREATMENT.

Crowned Snake *(Drysdalia coronata)* 40-65 cm

This is grey to pale or dark olive-brown on the body, the head being darker and with a black stripe forming a 'crown' that is most prominent on the nape. The lips are white or cream. A nocturnal species, it is found in south-western

Western Australia where it inhabits woodlands, heathlands and swamps. It feeds on frogs and lizards. Females give birth to from three to nine young in March or April. (This snake is sometimes referred to as *Elapognathus coronata*.) VENOMOUS BUT NOT REGARDED AS DANGEROUS.

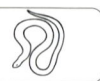

White-lipped Snake (*Drysdalia coronoides*) 35-50 cm

Two colour variations of the White-lipped Snake.

This snake has a conspicuous white stripe along the upper lips. The colour of the upperparts is extremely variable, ranging from grey, olive-green, and brown to reddish-brown: it may be yellow, cream or pink below. It occurs through south-eastern Australia and Tasmania, with a disjunct population in the north-eastern highlands of New South Wales. Frequenting woodland and dry sclerophyll forest habitats, especially where there is a tussock groundcover, it seems to prefer wetter areas within these associations. The main component of its diet is lizards, but frogs and reptile eggs are also taken. Females give birth to from two to ten young from February to April. (This snake is sometimes referred to as *Elapognathus coronoides*.)
VENOMOUS BUT NOT REGARDED AS DANGEROUS.

Masters' Snake *(Drysdalia mastersii)* 30-40 cm

The head is dark grey with a yellow or cream band on the nape and a narrow black stripe from the snout to the neck. The upperparts are yellow-brown to grey-brown, sometimes with a reddish tinge. The belly is yellow with dark flecks. Occurring in western Victoria, southern South Australia and south-eastern Western Australia, it is found in mallee and heathland in sandy areas. It is diurnal and feeds on lizards. The females give birth to two to three young. (This snake is sometimes referred to as *Elapognathus mastersii*.) VENOMOUS BUT NOT REGARDED AS DANGEROUS.

Bardick *(Echiopsis curta)* 40-70 cm

This snake has a stout body and a broad, depressed head that is distinct from the neck. It is grey, olive-brown or reddish-brown above, the sides of the head and neck being flecked with white: it is cream to grey below. Occurring in western Victoria and New South Wales, southern South Australia and south-western Western Australia, it is usually found in arid areas with mallee and *Triodia* hummock grass, woodlands or heath. Feeds on lizards, frogs and small mammals. Females produce from three to 14 young. VENOMOUS. A BITE MAY REQUIRE TREATMENT.

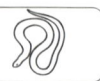

Red-naped Snake *(Furina diadema)* 35-45 cm

The red nape is completely enclosed by black.

This slender snake has a shiny black head and neck. There is a red bar on the nape, which is totally enclosed by the black neck colour. The body colour is reddish-brown with each scale dark-edged giving the snake a reticulated appearance. The eyes are small and black. Occurring in mid-eastern Queensland and most of New South Wales through to eastern South Australia, it is found in a wide range of habitats, often in association with termite or ant colonies. It feeds exclusively on small lizards and females produce a clutch of one to five eggs. VENOMOUS BUT NOT REGARDED AS DANGEROUS.

Orange-naped Snake *(Furina ornata)* 50-70 cm

Larger than the related Red-naped Snake, this species has a dark-brown to black head and neck, with an intervening orange band on the nape completely separating the dark head and neck colouration. The body is slender and pale- to dark-orange or reddish-brown in colour. The eyes are small and black. Occurring in Queensland, the Northern Territory, northern South Australia and Western Australia (except for the southern parts), it is found in woodlands, shrublands and hummock grasslands. This snake feeds on small lizards and females lay three to six eggs in a clutch. VENOMOUS BUT NOT REGARDED AS DANGEROUS.

Grey Snake *(Hemiaspis damelii)* 45-70 cm

Adults are grey to olive-grey above, often with a black spot at the base of each scale. In younger specimens the head is black but this colouration is reduced to a dark band on the nape in older specimens. The belly is cream with scattered grey spots. Occurring through inland New South Wales and south-eastern Queensland to the coast, it inhabits dry sclerophyll forest and woodlands in the vicinity of watercourses. It seems to concentrate its activity around dusk and feeds mainly on frogs, although lizards are also taken. From six to ten young are produced between January and March. VENOMOUS BUT NOT REGARDED AS DANGEROUS.

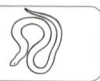

Marsh Snake *(Hemiaspis signata)* 50-90 cm

This snake is distinguished by two white to yellow streaks on the head, one from the eye to the neck, the other on the upper lips. The upper-body colour ranges from brown, olive-brown or dark olive-grey, to black or a pinkish-brown. It is usually dark grey to blackish below but some individuals have a salmon-coloured belly. Occurring in eastern Queensland and New South Wales to the south coast, it prefers moister habitats. The main food items are lizards and frogs, but it is also known to eat reptile eggs. From three to 20 young are produced in a litter. VENOMOUS BUT NOT REGARDED AS DANGEROUS.

Pale-headed Snake *(Hoplocephalus bitorquatus)* 50-100 cm

This snake is an overall brown or grey with a broad head that is distinct from the neck. There is a distinct white to grey band on the nape, and the head is spotted with black. The lips are cream with dark-grey to black bars. The belly is cream with darker flecks and the ventral scales are keeled. It occurs in eastern Australia on the coast, ranges and western slopes from central eastern New South Wales to Cape York Peninsula in Queensland. Arboreal and nocturnal, it prefers timbered areas, feeding on frogs and lizards but sometimes taking small mammals. From two to 11 young are born in late summer. It can be an aggressive snake if aroused. VENOMOUS. A BITE MAY REQUIRE TREATMENT.

Broad-headed Snake *(Hoplocephalus bungaroides)* 50-100 cm

This very distinctive snake is black on the back and sides, with yellow scales that usually form narrow irregular cross-bands. It is grey to grey-black below with the ventral scales keeled. The head is broad and distinct from the neck. It has a restricted range in New South Wales within a 250-kilometre radius of Sydney. Nocturnal by nature, it inhabits sandstone ridges and escarpments and feeds on lizards. From four to eight young are born in late summer. It is aggressive if aroused. VENOMOUS. A BITE MAY REQUIRE TREATMENT.

Stephens' Banded Snake *(Hoplocephalus stephensii)* 50-120 cm

This resembles the Broad-headed Snake in being dark-brown to black above with a series of narrow light cross-bands on the body and tail, but the bands are further apart. It has a broad black head which is distinct from the neck, with a brown crown and a cream patch on either side of the nape. The lips are barred with black and cream. It is cream below with black blotches and the ventral scales are keeled. It occurs in coastal areas from central New South Wales to south-eastern Queensland, and is found in wet sclerophyll forest and rainforest. It feeds on lizards, small mammals and frogs, and is arboreal and nocturnal. From three to eight young are born in February or March. VENOMOUS. A BITE MAY REQUIRE TREATMENT.

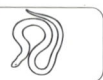

Krefft's Tiger Snake *(Notechis ater ater)* 90-120 cm

This species has three subspecies which are so distinct as to warrant separate treatment. Krefft's Tiger Snake has a robust body with a head distinct from the neck. It is greyish-black to black in colour, sometimes with paler cross-bands. Occurring in the Flinders Ranges and Mount Lofty Ranges of South Australia, it is found along creek margins where there is a thick understorey. It feeds on frogs, lizards, small mammals and nestling birds. From six to 15 young are produced in a litter. VENOMOUS AND DANGEROUS. CAPABLE OF INFLICTING A POTENTIALLY FATAL BITE.

Peninsular Tiger Snake *(Notechis ater niger)* 100-180 cm

This subspecies is usually black in colour, without any pattern (although juveniles may have whitish bands), and it has a blunt head and a robust body. Occurring on Kangaroo Island, the Sir Joseph Banks group of islands and the Eyre Peninsula in South Australia, it may be found in animal burrows, dense vegetation or under rock or timber. Its prey consists of small mammals, nestling birds and frogs. Some of the island forms feed almost exclusively on various sea-birds. From five to 20 young may be produced in a litter. VENOMOUS AND DANGEROUS. CAPABLE OF INFLICTING A POTENTIALLY FATAL BITE.

Tasmanian Tiger Snake *(Notechis ater humphreysi)* 100-160 cm

Not listed

There is considerable colour variation in this subspecies, body colour ranging from black, grey and chocolate-brown to tan, yellow or cream. Often there are lighter or darker crossbands. It occurs in Tasmania and on the Bass Strait islands. In Tasmania it is found from coastal heaths to highland forests. The Bass Strait island populations inhabit mutton-bird colonies. Food items comprise mammals, birds and frogs. On some of the Bass Strait islands it feeds exclusively on young mutton-birds. VENOMOUS AND DANGEROUS. CAPABLE OF INFLICTING A POTENTIALLY FATAL BITE.

Eastern Tiger Snake *(Notechis scutatus scutatus)* 120-200 cm

J. Cornish
G. E. Schmida

The Tiger Snake takes to the water readily.

The body of this subspecies is robust and the head is slightly distinct from the neck. The colour of the upperparts ranges from olive-grey, tan, and dark-brown to black (with or without the yellowish cross-bands which give the 'tiger' appearance). It occurs in south-eastern Queensland, eastern and southern New South Wales, Victoria and south-eastern South Australia. It has a preference for moist or swampy areas. Although usually shy, it performs an impressive threat display if cornered, hissing loudly and inflating the body. The neck and forebody are flattened to a considerable degree. It feeds mainly on mammals, frogs and nestling birds. Females give birth to between 14 and 80 young in summer. VENOMOUS AND DANGEROUS. CAPABLE OF INFLICTING A POTENTIALLY FATAL BITE.

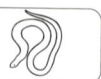

Western Tiger Snake *(Notechis scutatus occidentalis)* 100-160 cm

This tiger snake is brown, blackish-brown or black in colour with numerous yellow crossbands, although unbanded specimens do occur. The belly is yellow, becoming black towards the tail. It occurs in south-western Western Australia, where it is found in shrublands, woodlands and wet sclerophyll forest, usually in association with swamps or permanent watercourses. It feeds on mammals, lizards and frogs. Females give birth to from ten to 30 young, although as many as 90 have been recorded. VENOMOUS AND DANGEROUS. CAPABLE OF INFLICTING A POTENTIALLY FATAL BITE.

Inland Taipan *(Oxyuranus microlepidotus)* 180-250 cm

This taipan is olive-brown to dark brown above, many of the scales having a dark-brown or black edge, which give a speckled appearance or form obscure cross-bands. The head is often a glossy black. The belly is yellow, each scale having a dark edge. Occurring in south-western Queensland, south-eastern Northern Territory, north-eastern South Australia and western New South Wales, it is usually associated with the flat plains country, where it shelters in rat burrows or deep soil cracks. It feeds on mammals, particularly the Plague Rat *Rattus villosissimus*. Females lay 12 to 20 eggs in a clutch. Although placid and shy, it is nevertheless considered to be the most venomous land snake in the world. VENOMOUS AND DANGEROUS. CAPABLE OF INFLICTING A POTENTIALLY FATAL BITE.

Taipan *(Oxyuranus scutellatus)* 180-300 cm

G.E. Schmida

The scales on the neck and forebody of this species are weakly keeled and the head is long, deep and distinct from the neck. The eyes are brownish-red. The colour of the upperparts ranges from light to dark brown with the head often a paler colour, sometimes confined to the snout: the belly is cream to yellow. Occurring in coastal Australia from the Kimberley region of Western Australia to the New South Wales/Queensland border, it inhabits open savanna woodland and dry sclerophyll forest. It is swift and alert, with keen vision, and is rarely encountered in the wild. It feeds mainly on mammals but has been recorded as eating birds. If cornered, it defends itself aggressively and bites with great speed and accuracy. Female Taipans lay three to 25 eggs in a clutch. VENOMOUS AND DANGEROUS. CAPABLE OF INFLICTING A POTENTIALLY FATAL BITE.

Mulga Snake *(Pseudechis australis)* 150-270 cm

This is a robust snake with the head only slightly distinct from the neck. Colour of the upperparts ranges from coppery-brown, reddish-brown and olive-brown to almost black. The scales often have a darker edge which form a reticulated pattern. It occurs over most of continental Australia

J. Cornish

except for eastern New South Wales, most of Victoria and south-eastern South Australia, and southern Western Australia. It is diurnal in the southern part of its range but nocturnal in the more northern parts. It preys upon reptiles (including other snakes), mammals, frogs and birds. When provoked, it flattens the neck and forebody. Females lay from ten to 16 eggs. VENOMOUS AND DANGEROUS. CAPABLE OF INFLICTING A POTENTIALLY FATAL BITE.

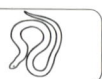

Spotted Mulga Snake *(Pseudechis butleri)* 100-160 cm

This elapid snake is black in colour, with cream- or yellow-centred scales grouped into irregular patches: the belly is yellow. The top of the head and the neck have few, if any, pale spots and the snout and side of the head are a reddish-brown. Occurring in the inland regions of southern Western Australia, it shelters in animal burrows, rock crevices or fallen timber. It feeds almost exclusively on lizards, but mammals are sometimes taken. The usual clutch is seven to 12 eggs. VENOMOUS AND DANGEROUS. CAPABLE OF INFLICTING A POTENTIALLY FATAL BITE.

Collett's Snake *(Pseudechis colletti)* 100-200 cm

This spectacular snake is rich-brown to black above, with numerous cross-bands of cream, pink or salmon: the lower sides are the same colour as the cross-bands. Occurring in central western Queensland, it inhabits the black-soil flood-plains, where it shelters in deep soil-cracks and fallen timber, becoming active after rain. It preys upon mammals and frogs. The female lays six to 14 eggs. VENOMOUS AND DANGEROUS AND CAPABLE OF INFLICTING A POTENTIALLY FATAL BITE.

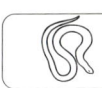

Blue-bellied Black Snake *(Pseudechis guttatus)* 100-195 cm

Spotted form of the Blue-bellied Black Snake.

Although usually black in colour, some animals may have a cream centre to each scale: others are cream in colour with no more than a black edge on each scale. The belly is bluish-grey. Young individuals are silver-grey with a darker head. This is a robust-bodied snake, found in north-eastern New South Wales and south-eastern Queensland. It feeds on frogs, mammals and lizards. Normally shy, it flattens the neck and forebody if provoked, hissing loudly. The clutch has an average of 12 eggs. VENOMOUS AND DANGEROUS. CAPABLE OF INFLICTING A POTENTIALLY FATAL BITE.

Red-bellied Black Snake *(Pseudechis porphyriacus)* 150-250 cm

This well-known snake is glossy black in colour with red, orange or pink on the lower sides, merging into a red belly. The snout is often a light brown. Occurring in eastern New South Wales, south-eastern Queensland, Victoria and south-eastern South Australia, it frequents well-watered areas, feeding on frogs, lizards and mammals. The female gives birth to from five to 40 young, which are actually born in individual membraneous sacs from which they emerge very shortly after birth. VENOMOUS AND DANGEROUS. CAPABLE OF INFLICTING A POTENTIALLY FATAL BITE.

Dugite *(Pseudonaja affinis)* 150-200 cm

This much-feared snake has a slender body and a small head which grades imperceptibly into the neck. Body colouration varies from yellowish-brown, pale or dark brown, to grey and almost black, and there is a scattering of black scales on the body. The head is sometimes paler than the body and the inside of the mouth is pink, the inside of the throat black. Occurring from south-western Western Australia to southern South Australia, it shows a preference for sandy areas and is found in coastal dunes, semi-arid woodlands and shrublands. It preys mainly upon mammals and lizards. Females lay 13 to 20 eggs in a clutch. VENOMOUS AND DANGEROUS. CAPABLE OF INFLICTING A POTENTIALLY FATAL BITE.

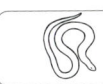

Western Brown Snake *(Pseudonaja nuchalis)* 100-150 cm

This snake has a slender body and a small head which is the same width as the neck. The mouth lining is blackish. There are at least nine forms of this species with a great number of colour patterns. The basic colour varies from olive-grey to dark brown or orange-brown. Some forms have a black head and nape, and others broad dark cross-bands with or without intervening narrower cross-bands; others have a reticulated pattern formed by dark-edged scales and some have a series of dark scales on the nape, which

Two of the variations of P. nuchalis.

may be in the shape of a chevron. It is a fast-moving diurnal snake, found through most of Australia except on the east coast, in south-eastern Victoria and in southern Western Australia. It feeds on mammals and lizards. The females lay 11 to 22 eggs. VENOMOUS AND DANGEROUS. CAPABLE OF INFLICTING A POTENTIALLY FATAL BITE.

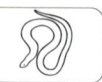

Speckled Brown Snake *(Pseudonaja guttata)* 100-140 cm

This species has a banded and an unbanded form. The body colour varies from pale fawn to orange-brown. In the unbanded form the edges of most of the scales are black, giving a speckled pattern. The banded form usually has nine to 18 broad dark bands across the body but these may be reduced to large dorsal blotches. The inside of the mouth is bluish-black in colour. Occurring in eastern Northern Territory, north-eastern South Australia and central Queensland, it inhabits tussock grasslands of the black-soil plains. It feeds on frogs, lizards and mammals. Females lay an average of six eggs. VENOMOUS AND DANGEROUS. CAPABLE OF INFLICTING A POTENTIALLY FATAL BITE.

Ingram's Brown Snake *(Pseudonaja ingrami)* 100-180 cm

The colour of this species ranges from a pale yellowish-brown to a blackish-brown, with the head blackish or a darker brown. The scales often have a darker edge. The belly is yellow to orange, with orange blotches arranged in parallel rows. The inside of the mouth is black. Occurring on the Barkly tableland of the Northern Territory and adjacent areas of Queensland, it inhabits black-soil plains where there is tussock grass cover. It feeds on mammals. Up to 15 eggs are laid per clutch. VENOMOUS AND DANGEROUS. CAPABLE OF INFLICTING A POTENTIALLY FATAL BITE.

Eastern Brown Snake *(Pseudonaja textilis)* 150-250 cm

A banded juvenile.

This brown snake has a slender body with a head that is barely distinct from the neck. Colour varies from light tan to dark-brown to blackish. The mouth lining is pink. In juveniles the top of the head is black and, in some regions, the body is strongly banded. Occurring in Queensland, New South Wales, Victoria and south-eastern South Australia, with isolated populations in the Northern Territory and Western Australia, it is diurnal and prefers drier areas. The species seems to have benefited from European settlement, being most abundant in agricultural areas, probably because of the mice that multiply there. It is aggressive if provoked, holding its forebody in a high S-shape and striking repeatedly. It feeds on mammals, lizards and frogs. The usual clutch is ten to 35 eggs. VENOMOUS AND DANGEROUS. CAPABLE OF INFLICTING A POTENTIALLY FATAL BITE.

Ringed Brown Snake *(Pseudonaja modesta)* 40-60 cm

The bands vary from 1 to 5 scales in width.

This relative of the Dugite has a slender body which is olive, tan or red-brown with four to 12 narrow black bands evenly spaced along the body. The bands may be obscure or absent in older individuals. The top of the head and the nape are dark-brown to black. The inside of the mouth is black. Distributed through the arid areas of Western Australia, Northern Territory, South Australia, Queensland and New South Wales, it often shelters in lizard burrows and its diet appears to consist of small skinks and geckos. The usual clutch is six to 11 eggs. VENOMOUS BUT NOT REGARDED AS DANGEROUS.

Eastern Carpentaria Snake *(Rhinoplocephalus boschmai)* 45-55 cm

The upperparts of this slender snake are lead-grey, light-tan, brown or dark brown in colour, becoming lighter on the sides. The head, which is often yellowish- to reddish-brown or the same colour as the back, is barely distinct from the neck and is depressed. The eyes are small.

Occurring in northern inland and eastern Queensland, this is a nocturnal species found in woodland habitats, where it shelters under fallen timber or ground debris. It feeds on small lizards. Females usually give birth to three to six young. (This snake is sometimes referred to as *Cryptophis boschmai*.) VENOMOUS BUT NOT REGARDED AS DANGEROUS.

Small-eyed Snake *(Rhinoplocephalus nigrescens)* 40-120 cm

This snake is a uniform shiny black or grey above and cream or pink below. It has a robust body and the head is distinct from the neck. It has noticeably small black eyes, hence the common name. Occurring in coastal eastern Australia, it is a nocturnal and secretive species

which shelters beneath rocks and sometimes beneath the loose bark of dead trees. It feeds on small lizards. Females usually give birth to two to eight young. (This snake is sometimes referred to as *Cryptophis nigrescens*.) VENOMOUS AND DANGEROUS. CAPABLE OF INFLICTING A POTENTIALLY FATAL BITE.

Black-striped Snake *(Rhinoplocephalus nigrostriatus)* 50-60 cm

This slender, large-eyed snake is pinkish- to dark reddish-brown above, each scale having a lighter edging. The top of the head is dark-brown to black and a distinct black vertebral stripe extends along the back to the tip of the tail. Occurring in coastal north-eastern Queensland, this nocturnal species inhabits woodlands and sclerophyll forests, sheltering beneath rocks, timber or ground debris. It feeds on small lizards. From two to six young are produced per litter. (This snake is sometimes referred to as *Cryptophis nigrostriatus*.)
VENOMOUS BUT NOT REGARDED AS DANGEROUS.

North-western Shovel-nosed Snake *(Simoselaps approximans)* 25-35 cm

The snout of this snake is upturned and shovel-shaped. Its colour varies from gray to dark brown with a broad black bar on the head and on the nape. There are numerous very narrow cream cross-bands which may be reduced to spots in some individuals and which do not form complete rings. It occurs in north-western Western Australia and is found on sandy plains and slopes. It is unusual in feeding exclusively on reptile eggs. The usual clutch is two to four eggs.
VENOMOUS BUT NOT REGARDED AS DANGEROUS.

Coral Snake *(Simoselaps australis)* 30-50 cm

This is pink to red above, with numerous cross-bands of cream-centred, dark-edged scales. There is a broad black bar across the head and another over the nape. The strong snout is shovel-shaped. Occurring in central New South Wales, south-eastern Queensland, south-eastern South Australia and western Victoria, this nocturnal burrowing snake is found in open woodlands and mallee where it favours red sandy areas. It feeds predominantly on reptile eggs but also eats small lizards. The usual clutch is four to six eggs. VENOMOUS BUT NOT REGARDED AS DANGEROUS.

Southern Desert Banded Snake *(Simoselaps bertholdi)* 30-35 cm

The upperparts of the body of this shovel-nosed snake are yellow to reddish with numerous dark or black cross-bands. The head is a mottled grey-brown or black, usually lighter on the snout. It occurs in southern Western Australia and over most of South Australia. A burrowing species, it feeds exclusively on small lizards. The usual clutch is one to eight eggs. VENOMOUS BUT NOT REGARDED AS DANGEROUS.

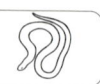

Black-naped Burrowing Snake *(Simoselaps bimaculatus)* 35-45 cm

This slender snake is pinkish- to reddish-brown above, each scale having a dark edging. There is a broad dark band across the head and another over the nape. This burrowing species occurs in southern Western Australia and south-western South Australia, where it is found in woodlands, shrublands and heath, sheltering under well-embedded timber or stumps. It feeds on lizards. The usual clutch is two to six eggs. VENOMOUS BUT NOT REGARDED AS DANGEROUS.

Black-striped Burrowing Snake *(Simoselaps calonotus)* 20-30 cm

This is Australia's smallest elapid snake. It has a cream-coloured body and each scale has pink or orange edges. The snout is tipped with black and there is a black band across the head and another over the nape, continued into a vertebral stripe of dark scales with white centres. Occurring along a narrow coastal strip of south-western Western Australia, it inhabits coastal heaths and shrublands, sheltering in soil beneath leaf litter, rocks or shrubs. It feeds on small lizards. The usual clutch is two to five eggs. VENOMOUS BUT NOT REGARDED AS DANGEROUS.

Narrow-banded Shovel-nosed Snake *(Simoselaps fasciolatus)* 35-40 cm

This is cream to reddish with numerous irregular, ragged-edged, blackish cross-bands and with scattered dark or reddish-spotted scales in the pale interspaces. There is a broad dark band on the head and another on the nape. Fossorial and nocturnal, it occurs in central Australia where it inhabits woodlands and shrublands, sheltering under stumps, timber and other ground debris. It feeds on lizards and their eggs. The average clutch is five eggs. VENOMOUS BUT NOT REGARDED AS DANGEROUS.

Southern Shovel-nosed Snake *(Simoselaps semifasciatus)* 30-40 cm

The rostral scale of this species has an upturned, sharply angular leading edge. The upperparts are light-brown to reddish with numerous dark-brown cross-bands and there is a broad black bar across the top of the head and another across the nape. Occurring in western and northern Australia it inhabits woodland and heathland. It is a nocturnal burrowing species that feeds exclusively on reptile eggs. The usual clutch is two to five eggs. VENOMOUS BUT NOT REGARDED AS DANGEROUS.

Little Whip Snake *(Suta flagellum)* 30-40 cm

This snake has small eyes and a depressed head. It is grey-brown to reddish-brown with each scale darker at the base. The top of the head and the nape are glossy black, this pigmentation being divided in front of the eye by a pale bar. The lips are cream or light tan. Occurring in Victoria, south-eastern New South Wales and south-eastern South Australia, it inhabits rock outcrops and woodlands where it feeds on lizards. Females give birth to three to five young. (This snake is sometimes referred to as *Rhinoplocephalus flagellum*.) VENOMOUS BUT NOT REGARDED AS DANGEROUS.

Monk Snake *(Suta monachus)* 40-50 cm

This slender snake has small eyes and a head that is clearly distinct from the neck. It is reddish-brown with a shiny black head and nape (the monk's 'hood'): the lips are white. Occurring in north-western South Australia and New South Wales, and south-western areas of the Northern Territory, it inhabits mulga and other arid shrublands, where it shelters under timber, in grass hummocks or in leaf litter. It feeds on small lizards. Females give birth to an average of three young. (This snake is sometimes referred to as *Rhinoplocephalus monachus*.) VENOMOUS BUT NOT REGARDED AS DANGEROUS.

Mallee Black-backed Snake *(Suta nigriceps)* 45-60 cm

Like the Monk Snake, this has a depressed head and small eyes. The head and nape are glossy black and there is a prominent black vertebral stripe from the nape to the tip of the tail. The rest of the back and sides is grey-brown or reddish-brown in colour. Occurring in southern Western Australia and South Australia, south-western New South Wales and north-western Victoria, it inhabits mallee and saltbush, where it shelters in leaf litter, spinifex hummocks and under ground debris. It preys on small lizards. Females give birth to three to six young. (This snake is sometimes referred to as *Rhinoplocephalus nigriceps*.) VENOMOUS BUT NOT REGARDED AS DANGEROUS.

Spotted Snake *(Suta punctata)* 40-55 cm

The colour of this snake ranges from yellowish-brown to reddish-brown, each scale usually having a darker spot or smudge. The head and neck have prominent dark blotches and the upper lips are white or pale cream. It is found in northern Australia (except eastern Queensland) where it favours red sandy loamy soil in woodlands and hummock grasslands. It feeds on small lizards. Females give birth to from two to five young. (This snake is sometimes referred to as *Rhinoplocephalus punctatus*.) VENOMOUS BUT NOT REGARDED AS DANGEROUS.

Spectacled Hooded Snake *(Suta spectabilis)* 30-45 cm

This snake is distinguished by a black 'hood' on the head and a pale patch in front of and behind each eye. The body is a greyish-brown to reddish-brown, usually with a dark spot at the base of each scale. It is widely distributed over southern central Queensland, central New South Wales, western Victoria, southern South Australia and south-eastern Western Australia and is found in a wide variety of habitats. It feeds on small lizards. Females give birth to from two to six young. (This snake is sometimes referred to as *Rhinoplocephalus spectabilis*.) VENOMOUS BUT NOT REGARDED AS DANGEROUS.

Curl Snake *(Suta suta)* 50-90 cm

The name of this snake refers to its defensive posture, where it curls its body into a series of coils and strikes out from this position. It has a relatively robust body and a broad flat head. The head and nape are dark-brown to black with a whitish to tan stripe running from the snout through the eye to the temple. The body colour can be any shade of brown, often with a dark tip to each scale. It is found in arid areas in Queensland, the Northern Territory, western Victoria and New South Wales, and South Australia and feeds on lizards, small mammals, frogs and reptile eggs. Females give birth to from two to seven young. (This snake is sometimes referred to as *Denisonia suta*.) VENOMOUS. A BITE MAY REQUIRE TREATMENT.

Rough-scaled Snake *(Tropidechis carinatus)* 75-100 cm

The scales on this snake are strongly keeled. It has a robust body and a head that is distinct from the neck. It is olive-green to dark brown, usually with narrow dark cross-bands. The belly is creamy-yellow to olive-green. There are 23 scale rows around mid-body. Occurring in coastal north-eastern New South Wales and south-eastern (and possibly north-eastern) Queensland, it inhabits rainforest and wet sclerophyll forest. Although essentially terrestrial, it will climb into low foliage to bask and forage. It is a shy snake but very aggressive when aroused. It feeds mainly on mammals and frogs but also takes birds and lizards. Females give birth to between five and 18 young. VENOMOUS AND DANGEROUS. CAPABLE OF INFLICTING A POTENTIALLY FATAL BITE.

Bandy-bandy *(Vermicella annulata)* 50-80 cm

This slender snake has a distinctive colour pattern of narrow white bands along the black body and tail. It is a nocturnal burrowing species which is found through northern and eastern Australia. It feeds predominantly, if not exclusively, on blind snakes. It has an unusual defensive display in which it raises its body into large loops. The usual clutch size is two to 13 eggs. VENOMOUS BUT NOT REGARDED AS DANGEROUS.

SEA-SNAKES Family HYDROPHIIDAE

These are completely aquatic marine snakes with vertically compressed paddle-shaped tails, and nostrils that are situated on the top of the snout with valvular flaps which close when the snake is submerged. All species give birth to their young underwater and do not have to come to land to lay eggs like other aquatic reptiles (for example the marine turtles and the sea kraits).

Horned Sea-snake (*Acalyptophis peronii*) 80-120 cm

This species has distinctive raised tubercles or spines on the scales above the eyes. The body scales are keeled and the forebody is slender, gradually thickening towards the tail. The colour is cream, grey or light brown with numerous dark cross-bands. There are 21 to 31 scale rows at mid-body. Occurring in the seas around northern Australia, it is usually found in inshore waters with a sandy or muddy bottom. It preys on fish. Females give birth to litters of four to ten young. VENOMOUS & DANGEROUS. CAPABLE OF INFLICTING A POTENTIALLY FATAL BITE.

Reef Shallows Sea-snake (*Aipysurus duboisii*) 70-120 cm

This sea-snake is cream to purplish-brown: a light edging to the scales creates a reticulated pattern. The body scales are smooth and there are 19 rows at mid-body. The belly scales have a small notch at the rear edge. Occurring in the seas around northern Australia, it is usually found in shallow reef waters. It feeds on small fish and eels. VENOMOUS & DANGEROUS. CAPABLE OF INFLICTING A POTENTIALLY FATAL BITE.

Olive Sea-snake *(Aipysurus laevis)* 150-200 cm

The body of this sea-snake is brown to purplish-brown, often with scattered creamy-white or spotted scales over the body. The tail is uniformly brown or white except for the dark-brown dorsal ridge. There are 21 to 25 scale rows at mid-body. It occurs in the seas around northern Australia (with occasional specimens appearing further south). It is usually found around coral reefs but also in the upper tidal areas of rivers. It feeds on fishes, shrimps and molluscs. Females give birth to litters of one to five young. VENOMOUS & DANGEROUS. CAPABLE OF INFLICTING A POTENTIALLY FATAL BITE.

Stokes' Sea-snake *(Astrotia stokesii)* 150-200 cm

This sea-snake has a large head which is distinct from the neck, and a very robust body. The small, paired belly scales form a distinctive median keel along the belly in adults. It is cream to almost black in colour, usually with darker cross-bands and spots. There are 45 to 63 scale rows at mid-body. Occurring in the seas around northern Australia, it is occasionally found further south. It inhabits coastal and reef waters where it feeds on fish. Females give birth to one to five young. VENOMOUS & DANGEROUS. CAPABLE OF INFLICTING A POTENTIALLY FATAL BITE.

Turtle-headed Sea-snake *(Emydocephalus annulatus)* 70-100 cm

The turtle-like appearance of the head is apparent here.

The head is not distinct from the neck in this species, which usually has a small blunt spine on the rostral scale at the front of the snout. The head bears a strong resemblance to that of a small marine turtle, hence the common name. The body scales are smooth and the belly scales have small tubercles and usually a keel. General body colour is dark-grey, brown or black, with or without numerous cream crossbands and scattered lighter and darker scales. There are 15 to 17 scale rows at mid-body. It occurs in the waters around northern Australia where it inhabits reef waters in particular. It feeds exclusively on fish eggs. VENOMOUS BUT NOT REGARDED AS DANGEROUS.

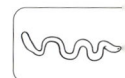

North-western Mangrove Sea-snake *(Ephalophis greyi)* 50-65 cm

The head is indistinctly separated from the neck in this species and the body scales are keeled, at least on the posterior portion of the body. It is cream to pale olive-brown with numerous dark-grey cross-bands or blotches. The head is speckled with dark grey. There are 19 to 21 scale rows at mid-body. Occurring in northern coastal Western Australia, it is found in mangroves and associated mud-flats. VENOMOUS & DANGEROUS. CAPABLE OF INFLICTING A POTENTIALLY FATAL BITE.

Black-ringed Mangrove Sea-snake *(Hydrelaps darwiniensis)* 40-60 cm

Smallest of the Australian sea-snakes, this species is cream or yellow with numerous dark cross-bands on the body and tail. The head is dark and speckled with yellow or cream. There are 25 to 29 scale rows at mid-body and these are smooth and not keeled. Occurring along the coast of northern Western Australia and the Northern Territory, it inhabits mud-flats associated with mangroves. VENOMOUS & DANGEROUS. PROBABLY CAPABLE OF INFLICTING A POTENTIALLY FATAL BITE.

Elegant Sea-snake *(Hydrophis elegans)* 150-200 cm

Active both by day and night, this sea-snake is characterized by a greatly elongated body and the deep and compressed tail fin in adults. It is pale-grey to brown with numerous dark cross-bands or blotches. In juveniles, the head is black and the cross-bands are very prominent. In adults the bands may be reduced to blotches or bars. There are 37 to 49 scale rows at mid-body. Found in the seas around northern Australia, but sometimes further south, it favours deeper waters than some of the other sea-snakes and feeds on eels. Up to 25 young are produced in a litter. VENOMOUS & DANGEROUS. CAPABLE OF INFLICTING A POTENTIALLY FATAL BITE.

Ornate Sea-snake *(Hydrophis ornatus)* 100-120 cm

This species is grey to bluish-grey with numerous broad dark cross-bands or blotches that become less distinct in older individuals. There are 39 to 59 scale rows at mid-body. It occurs in the waters around northern Australia and is found on coral reefs and in estuaries and inshore waters. VENOMOUS & DANGEROUS. CAPABLE OF INFLICTING A POTENTIALLY FATAL BITE.

Spine-bellied Sea-snake *(Lapemis hardwickii)* 100-120 cm

The scales on the underparts of this species become increasingly keeled as they approach the mid-line. In adult males, the keels of mid-ventral scales form small spines. The body is light to dark olive-grey, merging down the lower sides into a creamy belly colour. Some individuals, particularly juveniles, have a series of darker dorsal blotches joined at the vertebral line. There are 23 to 45 scale rows at mid-body. Occurring in the seas around northern Australia, it is found in a wide range of habitats in coastal waters, usually between six and 15 metres deep. Females give birth to litters of one to eight young. VENOMOUS & DANGEROUS. CAPABLE OF INFLICTING A POTENTIALLY FATAL BITE.

Yellow-bellied Sea-snake *(Pelamis platurus)* 60-80 cm

Very distinctive, this sea-snake is black or dark brown above and sharply yellow or light brown on the lower sides and belly. The head is long, narrow and distinct from the neck. The tail is yellowish and spotted with black. There are 47 to 69 scale rows at mid-body. It is usually encountered around the western, northern and eastern coastlines. Females give birth at sea to from two to six young. VENOMOUS & DANGEROUS. CAPABLE OF INFLICTING A POTENTIALLY FATAL BITE.

SEA KRAITS Family LATICAUDIDAE

Sea kraits are characterized by having smooth imbricate scales over the back and enlarged belly scales. They have laterally placed nostrils and the body has numerous black cross-bands. These snakes are partially terrestrial and, unlike the true sea-snakes which give birth to their young at sea, have to come ashore to lay their eggs. Two species are occasionally found in northern Australian waters.

Wide-faced Sea Krait *(Laticauda colubrina)* 100-140 cm

This marine snake is blue to bluish-grey with numerous black cross-bands which completely encircle the body. The head is black, except for the yellow snout and lips. There are 21 to 25 scale rows at mid-body. It is found occasionally along the Queensland coast. It feeds on gobies and eels and the females come ashore to lay their clutches of four to 20 eggs. VENOMOUS & DANGEROUS. CAPABLE OF INFLICTING A POTENTIALLY FATAL BITE.

CROCODILES Family CROCODYLIDAE

These largely aquatic primeval reptiles are represented in Australia by the two species described below. They are long and lizard-like (but, oddly enough, are more closely related to birds than to lizards!) and their bodies are covered with bony plates overlain by thick horny skin. Their limbs are short and their tails are long and muscular. They have several adaptations to an aquatic existence, for example their strongly webbed feet and their valvular nostrils.

Estuarine Crocodile *(Crocodylus porosus)* 300-500 cm

R. & A. Williams

This species has a broad snout with raised ridges running from in front of the eye to the centre of the snout. The body colour ranges from grey to almost black with some darker mottling. Occurring in the coastal regions of northern Western Australia, the Northern Territory and northern Queensland, it inhabits coastal rivers and swamps and is sometimes seen in the open sea; it also extends inland along the major rivers and into freshwater swamps. It feeds on fish, reptiles, mammals, birds and crustaceans. The female constructs a nest of vegetation and mud in which she deposits her clutch of up to 60 eggs. She stays near the nest during the incubation period and when the eggs hatch, she carries her diminutive offspring in her mouth to the water.

Freshwater Crocodile *(Crocodylus johnstoni)* 200-300 cm

The snout in this species is smooth, long and slender. Its body is grey to brown in colour with regular dark markings. Occurring in the permanent freshwater rivers and billabongs of northern Western Australia, the Northern Territory and northern Queensland, it is sometimes found in tidal rivers but only where the Estuarine Crocodile is not present. It feeds on fish, crustaceans, insects, frogs and any other animals small enough to be eaten. The female excavates a hole in a sandbank in which she lays her clutch of 12 to 20 eggs.

MARINE TURTLES Families CHELONIIDAE and DERMOCHELYIDAE

The feet of marine turtles have been modified to form flippers, especially the forefeet. All species spend their whole existence in the sea, except for the females which are obliged to come ashore to lay their eggs above the high-water mark on sandy beaches. They are protected by heavy armoured shells but, curiously enough, cannot withdraw their heads or limbs into the protection of the carapace. Six species occur in Australian waters.

Loggerhead Turtle *(Caretta caretta)* 80-150 cm

A. C. Robinson

The Loggerhead is dark brown in colour, sometimes with darker speckling. The head is disproportionately large, and the flippers relatively small. There are five costal shields on either side of the vertebral shields down the midline of the carapace. Occurring in tropical and warm temperate waters off the Australian coast, it feeds on molluscs, crustaceans and jellyfish. The female lays up to 100 round eggs in a hole excavated in the sand above the high-tide mark, usually in summer. After 60 to 80 days of incubation the hatchlings dig their way out of the nest and make for the sea, usually at night.

Green Turtle *(Chelonia mydas)* 80-150 cm

This turtle is greenish above with brown markings. There are four costal shields on either side of the carapace. The shields do not overlap and there is one claw on each front flipper. It occurs in shallow warm temperate and tropical coastal areas of Western Australia, Northern Territory and Queensland, particularly along the Great Barrier Reef. The female lays her clutch of large eggs (averaging 100) in a hole dug in the sand above the high-water mark, returning three to seven more times during the same season to deposit further batches of eggs. The Green Turtle is estimated to take from nine to 15 years to reach sexual maturity and there may be intervals of two to five years between nestings by one individual. Adults feed mainly on marine grasses although the young are carnivorous, feeding on jellyfish. Hatchlings are glossy black above and white below.

Leathery Turtle *(Dermochelys coriacea)* 120-220 cm

The carapace of this species looks as if it is covered with hard black rubber: it has seven longitudinal ridges and is sharply pointed at the rear. This is the largest living turtle and it occurs in tropical and temperate coastal waters around Australia, feeding mainly on jellyfish. (It often eats empty plastic bags floating in the water, mistaking these for jellyfish: such mistakes are fatal.) Females occasionally come ashore near Bundaberg in Queensland to lay their batches of 60 to 100 billiard-ball-sized eggs. They will return several times during the same season to lay more, the final total for each female being perhaps 400 eggs.

Hawksbill Turtle *(Eretmochelys imbricata)* 80-100 cm

The upper jaw in this species juts forward to form a beak and there are four costal shields on either side of the row of vertebral shields down the middle of the carapace. The body colour ranges from olive-green to brown with reddish-brown to black variegations. The shields overlap and the carapace is serrated at the rear. There are two claws on each front flipper. It occurs along the tropical and warm temperate coast of Australia from mid-western Western Australia to southern Queensland particularly on coral reefs, nesting on islands in the Gulf of Carpentaria and the Great Barrier Reef. Females lay up to 130 eggs at a time in holes dug into the sand above the high-water mark, repeating this two or three times at intervals of two to three weeks. Several years pass before individuals nest again. The diet consists mostly of molluscs, crustaceans, sponges and other relatively immobile animals, as well as jellyfish. The commercial 'tortoiseshell' is taken from the carapace of this species.

FRESHWATER TURTLES Families CHELIDAE AND CARETTOCHELYDIDAE

All members of these families are aquatic or semi-aquatic and have the nostrils situated at the tip of the snout as an adaptation to an aquatic existence. (They are used as a breathing 'snorkel' while the turtle remains hidden underwater.) The head and neck can be folded under the front edge of the carapace, except in the case of the Pitted-shelled Turtle below which is the only known member of the family Carettochelydidae. All have webbed paddle-like feet and all lay their eggs on land.

Pitted-shelled Turtle *(Carettochelys insculpta)* 45-75 cm

R.W.G. Jenkins

The sole member of its family, this freshwater turtle has a carapace covered with a soft pitted skin and paddle-shaped limbs similar to marine turtles. The nose is prominent, fleshy and pig-like. Grey to grey-brown in overall colour, it has a distinctive pale streak behind each eye. It is known from the Daly, Victoria and Alligator river-systems in the Northern Territory, where it feeds on pandanus fruit, snails and fishes. Females lay eight to 30 eggs in holes excavated in a sandbank; after developing fully, the young turtles lie dormant within the eggshells until the nest-site is flooded, when they emerge and enter the water.

Broad-shelled Snake-necked Turtle *(Chelodina expansa)* 40-50 cm

The head of this turtle is broad and depressed and its neck is longer than in any other long-necked turtle. It is brown to blackish-brown in colour, and whitish below with no dark colour on the seams of the plastron. There are four claws on each front foot. It is found in the Murray-Darling river-system of western New South Wales, Victoria and South Australia. It also extends through central south-eastern Queensland to the coast. It feeds on small fishes, crustaceans and aquatic insects. From nine to 17 eggs are laid by the females in autumn.

Eastern Snake-necked Turtle *(Chelodina longicollis)* 20-25 cm

The carapace of this turtle is brown to black and often covered in algae. A diagnostic feature is that the seams of the plastron are edged in black. Some hatchlings have orange or red plastrons. The long neck, which is covered with pointed tubercles, can be completely retracted beneath the front of the carapace. There are four claws on each front foot. Occurring through eastern Queensland, eastern New South Wales, Victoria and south-eastern South Australia, it is found in swamps, rivers and billabongs (and also often in man-made stock-tanks). It emits an extremely pungent and unpleasant odour when alarmed. It feeds upon a wide variety of aquatic creatures such as frogs, tadpoles, small fish and crustaceans. The female lays up to 24 eggs in a hole excavated in the bank above the waterline of her chosen habitat.

Northern Snake-necked Turtle *(Chelodina rugosa)* 30-40 cm

This species' carapace is dark-brown to black, often with darker flecks or blotches. The plastron is whitish. The head is very broad and depressed. There are four claws on each front foot. Occurring across the northern parts of Western Australia, Northern Territory and Queensland, it is found in slow-flowing rivers, billabongs and swamps. It does not produce an offensive odour when threatened, as does *Chelodina longicollis* (page 77). It feeds on frogs, tadpoles, fishes and crustaceans. Females lay 12 to 14 eggs which hatch with the onset of the wet season.

Northern Snapping Turtle *(Elseya dentata)* 25-35 cm

The carapace of this species, which is brown to dark brown with some darker blotches, is distinctly wider at the rear than at the front. Juveniles and young adults have serrations along the hind edge of the carapace. The nose is prominent and the neck is covered with small rounded tubercles. There are five claws on each forefoot. Adults often develop enormously big heads. Occurring in northern Western Australia, the northern regions of the Northern Territory and northern Queensland, it may be found in the large river-systems and lagoons. It feeds on molluscs, fish and carrion. Females lay about five eggs per clutch.

Krefft's River Turtle *(Emydura krefftii)* 20-25 cm

The carapace is pale brown to almost black, the plastron is bluish-green, and there is a yellow or blue-green stripe extending back from the eye. The carapace has a serrated hind edge in juveniles. There are five claws on each forefoot. This turtle is found in the larger rivers and water-holes in eastern and north-eastern Queensland and also on Fraser Island. It feeds on crustaceans, insects, molluscs and vegetation. From four to 16 eggs are laid in each clutch.

Murray Short-necked Tortoise *(Emydura macquarii)* 20-30 cm

The carapace of this turtle ranges from light to dark brown in colour and the plastron is white or pale yellow. The forefeet have five claws each. There is a yellow stripe running from beneath the chin along the neck. Although it occurs generally in the Murray-Darling river-system and associated tributaries in western New South Wales, southern Queensland, northern Victoria and south-eastern South Australia, it is found mainly in the larger rivers and lakes. It feeds on aquatic vegetation, crustaceans, fishes and tadpoles. The female lays up to 25 eggs in a hole dug into the river-bank.

GECKOS Family GEKKONIDAE

The majority of geckos are essentially nocturnal and all are capable of vocalization. They have well-developed fingers and toes, and a tail that is easily cast off in emergencies (to be regrown later). They have large prominent eyes and small granular non-shiny scales that do not overlap. In many species the digits end in an expanded pad that aids climbing on vertical and even overhanging surfaces. All Australian species lay eggs.

Marbled Southern Gecko *(Christinus marmoratus)* 7-14 cm

The colour ranges from grey to light pink-brown, with dark-brown or black reticulations over the head, body and limbs. Juveniles often have a row of red or orange spots down the tail. Occurring across the southern parts of Western Australia, South Australia and New South Wales and the central and northern parts of Victoria, it is found on standing trees with crevices or peeling bark, as well as in rock crevices and exfoliations. One or two eggs are laid in a communal nest (30 eggs have been found at one site) in November or December. Females may lay more than once in a season. This is the most southerly of the Australian geckos and is generally found in cooler, moister areas. (This gecko is sometimes referred to as *Phyllodactylus marmoratus*.)

Clawless Gecko *(Crenadactylus ocellatus)* 6-7.5 cm

H. Ehmann

The body colour is grey to brown with four or five darker stripes extending the length of the body: one commences at the snout and continues through the eye along the side of the body. Some individuals have numerous small pale ocelli ('eye-spots') on the back and sides.

Occurring throughout most of Western Australia and the Northern Territory, the western areas of Queensland and the northern parts of South Australia, it prefers rocky, stony areas with spinifex, hummock grasses or heathlands. Located within grass clumps or dense bushes, it forages for invertebrates within this cover. It is the smallest of the geckos in Australia and lacks claws. Two eggs are laid, usually in November or December.

Northern Spiny-tailed Gecko *(Diplodactylus ciliaris)* 10-15 cm

This extremely variable gecko ranges in colour from a pale grey to dark brown. There may be almost no pattern, a few scattered black and orange scales, or mottled areas of orange, white or brown. There are several prominent spines on the eyebrows and two rows of enlarged spines running along the back and tail. Occurring through the northern parts of Western Australia and South Australia, the Northern Territory, western Queensland and north-western New South Wales, it is usually arboreal and found in woodland but in some areas it inhabits spinifex clumps and low shrubs. It eats a wide range of insects. It is able to squirt an irritating sticky fluid from the spines on its tail accurately for up to 30 cm to deter a predator. Two eggs are laid per clutch.

Jewelled Gecko *(Diplodactylus elderi)* 7.5-9 cm

Dark-brown or grey to black with small dark-edged white spots, this gecko occurs in south-western Queensland, the southern parts of the Northern Territory, south-western New South Wales, most of South Australia and the interior and north-western parts of Western Australia. It is found only in spinifex clumps where it feeds at night on termites and other invertebrates. It has the ability to exude an unpalatable sticky substance from tubercles on the tail: this acts as a deterrent to predators. Two eggs are laid in January or February and the young hatch in March.

Fine-faced Gecko *(Diplodactylus pulcher)* 8-10 cm

This largely terrestrial gecko is light-brown to reddish-brown above with a pale crown on the head and usually has a series of pale dark-edged blotches from the nape on to the tail. Some individuals have a continuous vertebral stripe from the head to the tail. Found in central and southern Western Australia and the southern areas of South Australia, it favours acacia and mallee shrublands, sheltering beneath timber and rocks or in spider burrows. It is nocturnal and feeds almost exclusively on termites. A clutch of two eggs is laid in October and subsequent clutches may be produced in the same season.

Eastern Stone Gecko *(Diplodactylus vittatus)* 8-11 cm

The body colour is grey to brown with a pale zigzag stripe running down the middle of the back from the head on to the tail. There are scattered spots on the sides. Found in south-eastern Queensland, most of New South Wales except the far western and south-eastern corners, northern Victoria and south-eastern South Australia, it is terrestrial and usually spends the day under rock or timber. It prefers rubble-strewn areas with rock outcrops but is also found on red-soil plains. It feeds at night on spiders and insects. The female lays two eggs per clutch.

Eastern Spiny-tailed Gecko *(Diplodactylus williamsi)* 9-13 cm

This diurnal gecko is pale to dark grey with scattered black spots. Four rows of tubercles on the back are replaced on the tail by larger spiny light-orange tubercles. The tongue and inside of the mouth are dark purple. Occurring on the north-eastern coast of Queensland through the interior to north-eastern New South Wales and far north-western Victoria and adjacent areas in South Australia, it inhabits woodland and open forest where it shelters under loose bark on trees or in hollows in trees and stumps. It is arboreal and sometimes active during the day when it probably feeds on insects. It is able to squirt an irritant sticky fluid from the tubercles on the somewhat prehensile tail to deter predators (including humans!).

Variegated Dtella *(Gehyra variegata)* 7-12 cm

The colour of this gecko ranges from grey to grey-brown, with a chequered pattern of dark and pale blotches and marblings; there are several dark lines on the side of the head. Occurring in Western Australia, South Australia, the southern parts of the Northern Territory, northern Victoria, New South Wales and western Queensland, it is found in trees or, in some areas, under rock exfoliations. It feeds on a wide range of invertebrates, taking them under shelter during the day and in the open at night. One egg is laid per clutch with at least two clutches in a season. Communal egg-laying has been recorded.

Bynoe's Prickly Gecko *(Heteronotia binoei)* 9-12 cm

The colour of this widespread terrestrial gecko is extremely variable but basically there is a brown background with black markings that may form a pattern of cross-bands. The body is covered with numerous keeled scales which give it a prickly feel. Occurring throughout Western Australia (except the south-western corner), the Northern Territory, South Australia, Queensland, western New South Wales and north-western Victoria, it is found in most habitats and shelters under ground debris and fallen timber and in rock crevices. It is nocturnal and feeds on insects and other invertebrates. It is an extremely widespread gecko and is locally abundant. Several populations consist only of females which are able to reproduce without mating; this phenomenon is known as parthenogenesis. Two hard-shelled eggs are laid per clutch and communal egg-laying has been recorded.

Rough Knob-tailed Gecko *(Nephrurus asper)* 9-14 cm

As in all members of its genus, this gecko has a disproportionately large head and a very small tail which ends in a soft bulbous knob. It is pale-gray to reddish-brown, with transverse rows of pale spots alternating with blackish cross-bands. The back, sides and limbs have large tubercles surrounded by smaller rounded scales, making it prickly to the touch. Occurring in the northern areas of Western Australia, the Northern Territory and the adjacent areas of far-northern South Australia, and the western and northern regions of Queensland, it inhabits woodland and is usually found in rocky areas. It is nocturnal and shelters by day in crevices, or under rock slabs or fallen timber; it emerges at night to feed on spiders, cockroaches and other large invertebrates, as well as on other geckos. Females lay clutches of two eggs.

Three-lined Knob-tailed Gecko *(Nephrurus levis)* 9-15 cm

Red-brown to purple-brown above, this gecko has three pale-yellowish curving cross-bars, one across the back of the head, another across the neck and a third between the forelimbs. There are a number of small tubercles on the body and tail which may form bars or rows. The head is large and the short carrot-shaped tail ends in a bulbous knob. Occurring on the west coast and interior of Western Australia, through the southern areas of the Northern Territory and the interior of South Australia to north-western New South Wales and the south-western and western parts of Queensland, it is found in open woodland, shrublands and spinifex areas where it shelters by day in a burrow. It feeds on spiders, cockroaches and other invertebrates as well as on smaller geckos. It often engages in slow tail-waving when stalking prey and emits a squeaky bark when alarmed.

Marbled Velvet Gecko (*Oedura marmorata*) 12-18 cm

Juvenile showing prominent bands.

Brown to purplish-brown in background colour, this gecko typically has five or six broad yellow or cream transverse bands, sometimes with scattered brownish blotches between the bands. In older individuals the pattern may break down into blotches with no bands. A pale stripe along the lips usually merges with the first transverse band. Juveniles are strongly banded with no blotching. The tail is long and slender in some populations, short and thick and very depressed in others. Occurring through central and northern parts of Western Australia, much of the Northern Territory, northern South Australia, northwestern New South Wales and central and western Queensland, it is found in woodland and in rocky outcrops. It shelters by day under bark on dead trees, in holes and crevices in trees, and in rock crevices and exfoliations. It feeds on insects, smaller geckos and skinks. The clutch consists of two eggs.

Zig-zag Velvet Gecko *(Oedura rhombifer)* 8-11 cm

This largely arboreal gecko is grey to brown in colour with a prominent broad paler band running down the middle of the back and along the tail. This band has a ragged edge which has given rise to the common name. On the neck, the band divides into two broad branches which run on either side of the head. The sides of the body and the limbs have darker and paler flecks. The tail is long and relatively slender. Occurring in the northern areas of Western Australia and the Northern Territory, and northern and eastern Queensland, it inhabits woodland and open forest, where it shelters beneath loose bark on trees and in ground debris. It is occasionally found in rock outcrops and around buildings and it feeds on small insects.

Southern Leaf-tailed Gecko *(Phyllurus platurus)* 11-15 cm

In most leaf-tailed geckos, the tail is flat and broad and remarkably leaf-like. This species is grey to rich reddish-brown, mottled with darker spots over the entire surface. There are numerous low pointed tubercles on the back and a few low conical ones on the sides. Regenerated tails are smooth, lacking the tubercles found on original tails. Occurring in New South Wales in the central coastal areas and adjacent ranges, it inhabits deep crevices in sandstone outcrops where it forages within the rock crevices and caves in the sandstone. It is also found sometimes in little-used suburban buildings. It preys on spiders, cockroaches, crickets and similar invertebrates. When disturbed it emits a quite audible bark and gapes the mouth threateningly. It may also move its tail slowly from side to side during this defensive display. Two eggs are laid within rock crevices during November and December. Females practise communal egg-laying and will utilize the same site over many years.

Leaf-tailed Gecko *(Saltuarius swaini)* 14-20 cm

The tail of this species is remarkably wide and flattened. Light-grey to brown to olive-green above, it usually has several lighter transverse patterns from the head to the tail. The remaining areas are patterned with brown flecks and reticulations, producing a 'lichen' effect. There are scattered large pointed tubercles on the back and on the sides. Together with its ability to change colour to match its background, these features enable the gecko to blend in perfectly with the surfaces on which it forages. Occurring in south-eastern Queensland and north-eastern New South Wales, it inhabits rainforest and wet sclerophyll forest (except for a rock-dwelling form which occurs in dry sclerophyll forest and shelters in crevices or under exfoliations). The forest form shelters in hollows and under peeling bark on trees. Nocturnal and arboreal, it forages on both large trees and saplings at night. Two eggs are laid in November or December.

Thick-tailed Gecko *(Nephrurus milii)* 8-13 cm

The tail of this nocturnal gecko is carrot-shaped and the legs are long and slender. The upper surface of the body ranges in colour from very dark purple-black to dark reddish-brown. White or yellow spots form irregular rows across the back but more distinctive lines on the back of the head and nape. The head and eyes are large. Occurring in southern Queensland, New South Wales except the far south-eastern part, northern Victoria, southern South Australia, southern Western Australia and the far south of the Northern Territory, it is found in a wide range of habitats including woodland, sclerophyll forest and arid shrubland. It favours rocky areas where it shelters by day in crevices or under loose rock. In the more arid areas it utilizes the burrows of other animals for shelter. Aggressive if disturbed, it raises its body off the ground and makes a barking or coughing sound. It preys on insects and spiders. Two eggs make up the clutch and communal nesting is known to occur.

PYGOPODIDS Family PYGOPODIDAE

These lizards are snake- or worm-like with no trace of forelimbs. The hindlimbs are represented by, at most, scaly flaps. Most of the species have external ear openings and all have a broad fleshy tongue with a notch at the tip. The tail is long and easily cast off. All species are egg-laying and vocalize.

Burton's Snake-lizard *(Lialis burtonis)* 30-60 cm

Colour pattern variations.

A distinguishing feature of this species is the wedge-shaped head, unique among Australian reptiles. The basic colour varies from cream to deep brown, without any patterning, or with stripes and blotches. The tail is equal in length to, or longer than the robust body. The ear-opening is prominent and there is a small 'hindlimb' flap. Widely distributed in all mainland states, it is found in most habitats except rainforest, high-altitude regions of the Great Dividing Range and desert extremes. It feeds exclusively on other reptiles, particularly skinks, ambushing these by lying in wait under cover until they come within range. Although active by day and night, it is most likely to be seen moving about in the early evening. The clutch consists of two eggs and females probably lay more than once in a season. Communal egg-laying has been recorded.

Javelin Lizard *(Aclys concinna)* 30-50 cm

This very slender legless lizard has an elongate pointed snout. The 'hindlimb' flaps are well developed and the tail is about four times the length of the body. It is grey with narrow darker longitudinal stripes along the back and sides of the body and tail.
Occurring along the coast of southern Western Australia, it inhabits coastal heaths and woodlands. It has been seen to bask in elevated sites in bushes. It feeds on insects and spiders, but also eats some plant material.

Mallee Worm-lizard *(Aprasia inaurita)* 12-16 cm

This burrowing lizard is yellowish-brown to greyish-brown on the body: the head and neck are often a reddish-brown. The tail is a rich reddish-orange. Occurring in far south-eastern Western Australia, southern South Australia, north-eastern Victoria and south-western New
South Wales, it it is found in mallee woodland and coastal dune vegetation. The tail is shorter than the body. It has been seen to eat small black ants.

Many-lined Delma *(Delma impar)* 18-25 cm

This very attractive lizard is light grey-brown in colour usually with a series of darker, punctuated, longitudinal stripes, particularly along the side of the body and tail. The tail is two to three times the length of the body. Occurring in Victoria, south-eastern New South Wales and south-eastern South Australia, it inhabits grassy plains with little or no tree cover. It has been found under rocks and it sometimes utilizes spider burrows as shelter sites. It feeds on insects, spiders and moth larvae, ambushing these from shelter. Females lay their eggs in chambers at the base of their burrows.

Sharp-snouted Delma *(Delma nasuta)* 30-45 cm

A feature of this species is the long, pointed snout. It is olive-brown to deep brown, usually with a darker reticulated pattern formed by a dark edge or spot on each scale. The tail is about four times as long as the body. Occur-

ring in the northern half of Western Australia, the Northern Territory, western Queensland and the far north of South Australia, it is usually found in spinifex clumps in arid areas, where it feeds on insects and spiders. A species very similar in appearance is the Unbanded Delma *(Delma butleri)* which is found in southern areas of Western Australia, the interior of South Australia and the western areas of New South Wales.

Excitable Delma *(Delma tincta)* 20-35 cm

Photo: S. Swanson

The name of this species refers to its forceful writhing ('flick-leaping') when disturbed. It is grey to grey-brown, almost always with dark bands on the head (although these may be obscure in older animals). The tail is two to three time the length of the body. Occurring throughout Western Australia, Northern Territory, Queensland, the northern parts of South Australia and New South Wales, it usually inhabits open woodland or spinifex grasslands, where it shelters beneath surface debris or within a tussock. Known to feed on spiders and probably on insects, it conceals itself in loose soil or litter with only the head protruding, presumably to ambush prey. Females lay their two-egg clutches in spring.

Brigalow Scaly-foot *(Paradelma orientalis)* 30-40 cm

Photo: R. W. G. Jenkins

The only member of its genus, this rather robust pygopodid has a blunt rounded snout, prominent ear-openings and glossy scales. It is grey to brown in colour with a black bar on the neck behind a wide creamy-yellow band on the nape. The tail is twice the length of the body. It occurs in the Great Dividing Range and western slopes of central eastern Queensland where it is found in brigalow scrub and woodlands.

Slender Slider (*Pletholax gracilis*) 25-35 cm

The very slender body and tail distinguish this lizard, which also has keeled scales. The tail is three to four times the length of the body, which is grey with much darker longitudinal lines extending from the head to the tip of the tail. The throat and lips are bright yellow.

It occurs in the coastal regions of south-western Western Australia where it is found in dense coastal vegetation. It feeds on nectar, insects and spiders and the females lay clutches of two eggs in spring.

Southern Scaly-foot (*Pygopus lepidopodus*) 30-70 cm

This species mimics the behaviour of a venomous snake when it is threatened, by raising its head and forebody high off the ground and flattening the neck. Its dorsal scales are heavily keeled and not at all glossy, giving it a matt finish. It is grey to reddish-brown and may either lack markings or have a series of black dashes extending in rows down the length of the body and on to the tail. There are usually some dark bars on the lips and further oblique bars on the side of the neck. The tail is twice the length of the body and there is a large 'hindlimb' flap.

Occurring in the southern parts of Western Australia and South Australia, through Victoria and the eastern areas of New South Wales to the south-eastern areas of Queensland, it may be found in habitats ranging from dry sclerophyll forest and woodland to heath and mallee. Normally active during the day, it becomes nocturnal in hot weather. Otherwise it shelters beneath ground litter and in grass tussocks. It feeds mainly on spiders. The species engages in communal egg-laying and two eggs are laid per clutch.

Hooded Scaly-foot *(Pygopus nigriceps)* 25-45 cm

The basic body colour of this pygopodid ranges from grey to red-brown and there is usually a pattern consisting of reticulations created by dark-edged scales, or scattered lighter and darker scales forming indistinct longitudinal patterns. The head has a dark 'hood' made up of

two blackish bands which may merge into one or become obscure in older specimens. The scales are smooth or only faintly keeled. The tail is twice the length of the body and the 'hindlimb' flaps are conspicuous. Occurring throughout most of mainland Australia except the coast eastward of the Great Dividing Range and the southern coastal and highland regions, it occupies a variety of habitats from black-soil plains to red-sand ridges, with vegetation ranging from hummock grasses to woodland. It shelters under ground debris, in burrows and cracks in the soil and within tussocks or dense ground vegetation, tending to be crepuscular and nocturnal through most of its range. It feeds on insects and spiders. Females lay two eggs per clutch.

MONITORS or GOANNAS Family VARANIDAE

Monitors or goannas include among their number the largest lizard in the world, the Komodo 'dragon' of Indonesia. Australia has 25 described species of monitor, more than 80 per cent of the world's total. One characteristic of monitors is the long, deeply forked tongue which is constantly flicked in and out. Their skins are loose-fitting, dull and rough and their heads are long and flat with long necks. The tail is long and slender and the limbs are well developed with sharp claws. All species lay eggs.

Ocellate Ridge-tailed Monitor *(Varanus acanthurus)* 50-75 cm

The body colour is black, dark-brown or reddish-brown with numerous red to cream ocelli ('eye-spots') on the back and flanks. The tail is brown with rings of light-brown or yellow scales and has a vertebral and two lateral ridges running the whole length. The scales on the tail are spiny, helping the animal to anchor itself in rock crevices as a defence against predators. Found in the northern half of Western Australia, most of the Northern Territory, the far northern areas of South Australia and the far western areas of Queensland, it inhabits rocky outcrops in woodlands with hummock grasses. It is found in rock crevices, termite-mounds or burrows beneath rocks and logs. It feeds on grasshoppers, cockroaches, beetles and lizards. Females lay from four to eight eggs in a burrow.

Short-tailed Pygmy Monitor *(Varanus brevicauda)* 20-25 cm

R. W. G. Jenkins

This species is creamy-brown to red-brown in colour, with darker and paler flecks over the body, limbs and tail. It is the smallest monitor in the world and attains a total length of 25 cm. As the name suggests, the tail is proportionally short and about the same length as the head and body. Occurring in the northern areas of Western Australia, southern Northern Territory, western Queensland and north-western South Australia, it may be found in sandy desert areas or stony plains, particularly with spinifex. It feeds on grasshoppers, cockroaches and beetles as well as reptile eggs. It shelters in burrows excavated beneath spinifex clumps or rocks and is very wary, spending much of its time under cover. There are two to three eggs in a clutch.

Rusty Desert Monitor *(Varanus eremius)* 35-45 cm

A. Greer

This goanna is reddish-brown with numerous irregular dark-brown to black flecks and smaller pale spots. The tail has alternating cream and dark-brown longitudinal stripes and there is a black stripe from the snout to the eye. The scales on the tail are keeled. Occurring in central Western Australia, southern Northern Territory and northern South Australia, it inhabits the sandy desert areas in shrublands and hummock grasses. It is alert and quickly retreats to its burrow when disturbed. It feeds on insects and small lizards. There are from three to six eggs in a clutch.

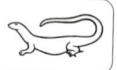

Perentie *(Varanus giganteus)* 150-250 cm

Perentine with throat pouch distended.

This is Australia's largest lizard. The body is cream and very densely speckled with dark brown. There are transverse rows of large yellowish blotches edged in dark brown across the body and tail, and a prominent reticulated pattern on the neck, side of the head and throat. Occurring in western Queensland, northern South Australia, the southern parts of the Northern Territory and the adjacent interior of Western Australia through to the north-west coast, it prefers rocky outcrops and gorges but forages over a wide area and may also be seen on the sandy plains. It shelters in rock crevices and burrows. It feeds on reptiles, small mammals, birds and carrion. It can run swiftly either on all four legs or on its hindlegs. When alarmed, an adult may stand up on its hindlegs and distend the throat pouch, emitting a loud hissing noise. The female lays six to 13 eggs in a burrow which she then fills in. The young are brightly coloured replicas of the adults.

Pygmy Mulga Monitor *(Varanus gilleni)* 25-35 cm

The body of this small monitor is depressed and the tail is keeled. It is reddish-brown to grey-brown in colour with narrow darker bands extending on to the tail. The latter half of the tail has five narrow dark longitudinal stripes. Occurring in central and western Australia, it inhabits arid areas, in particular those with mulga. It is arboreal and shelters under loose bark or in tree hollows. It feeds on lizards and insects. The female lays four to seven eggs.

Sand Monitor *(Varanus gouldii)* 100-160 cm

Head and forebody.

The colour of this monitor ranges from cream to dark brown with lighter and darker spots that form irregular transverse bands or ocelli ('eye-spots'). There is a black stripe behind the eye. The tail is banded except for the tip which is cream or yellow without any markings. It occurs throughout most of Australia apart from Tasmania, the extreme south and some parts of the east coast, preferring sandy soil and an open habitat. A terrestrial species, it shelters in burrows which it digs into the sandy soil. It also utilizes the burrows of other animals, particularly rabbit warrens. In its active foraging, the Sand Monitor feeds upon anything it can dig out or find, particularly lizards, spiders, reptile eggs, turtle eggs and carrion. The female lays three to 11 eggs in a burrow which is then filled in and concealed.

Mertens' Water Monitor *(Varanus mertensi)* 75-110 cm

Head and forebody.

This widespread and familiar aquatic lizard is olive-grey to brown in colour with numerous small pale spots on the back, sides and limbs. The tail is markedly compressed laterally to assist in swimming and has a distinct vertebral keel. Occurring in the northern areas of Western Australia, the Northern Territory and Queensland, it is found in the vicinity of swamps, lagoons and rivers. It is quite aquatic and is seldom found far from water. It is usually seen basking on branches overhanging the water or on logs and rocks. If alarmed, it dives into the water and remains submerged for a considerable period. It feeds on fishes, frogs, crabs and insects. The female lays from ten to 14 eggs in a burrow which is then filled in.

Black-tailed Monitor *(Varanus tristis)* 65-80 cm

Overall, this monitor is grey, brown or black with numerous cream to white ocelli, each with a dark centre. The head and neck are black in some populations. The tail is long, slender and keeled. It occurs through most of Western Australia, the Northern Territory, northern South Australia, Queensland and north-western New South Wales in a variety of habitats. It is mainly arboreal but may be found in rock crevices in some areas and is sometimes seen on roads, pressed to the ground with the tail arched in a high curve over the body. It feeds on lizards, insects, birds and birds' eggs. Females lay five to 17 eggs in a clutch.

Yellow-spotted Monitor *(Varanus panoptes)* 100-160 cm

This monitor is dark-brown to reddish-brown with a pattern of alternate transverse rows of dark and light spots, a pattern which extends on to the tail, the distal half of which is cream or light brown with darker bands. There is a

dark band through the eye and on to the neck. Occurring in the Kimberley region of Western Australia and the northern area of that State, the Northern Territory and northern and central Queensland, it is terrestrial and found in woodland and grasslands. It shelters in a self-made burrow excavated beneath rocks or bushes and feeds on insects, reptiles, small mammals and carrion. The species was for many years confused with the Sand Monitor, which is similar in size. Females lay from six to 13 eggs in summer.

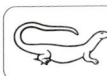

Heath Monitor *(Varanus rosenbergi)* 90-150 cm

The upperparts of this species are dark grey, with small yellow or white spots. Superimposed on this pattern are 12 to 15 blackish bands across the neck and body and a series of yellow and dark-brown bands along the tail. Occurring in the southern parts of Western Australia and South Australia, north-western Victoria, and with isolated populations in central coastal New South Wales, it is found in sclerophyll woodlands and heathlands where it favours sandy soil. It is a terrestrial species which forages for reptiles, insects, small mammals, birds and their eggs, and carrion. It shelters in a self-made burrow. The female lays up to 12 eggs in a cavity excavated into a termite-mound. The termites seal off the hole and provide a secure and 'air-conditioned' environment for the eggs to incubate.

Spencer's Monitor *(Varanus spenceri)* 90-120 cm

This monitor is light grey-brown with scattered darker and lighter spots and a series of irregular light-grey cross-bands on the body and tail. There are three to four V-shaped bands on the neck and shoulders. The tail is about the same length as the head and body. Occurring in north-western Queensland and eastern Northern Territory, it is found on black-soil plains with associated tussock grasses. It lives in a burrow or deep soil crack and feeds on lizards, small mammals and insects. The female lays ten to 30 eggs in a clutch.

Storr's Monitor *(Varanus storri)* 30-45 cm

The upperparts of this species are pale to dark-brown or reddish-brown with obscure darker reticulations or ocelli. The thick and spiny tail is spotted with black and brown. Occurring in north-western Queensland, and the Northern Territory through to northern Western Australia, it is found in crevices in rocky areas with grass cover or in burrows beneath rocks. Feeds on insects and small lizards. From two to seven eggs are laid.

Lace Monitor *(Varanus varius)* 100-200 cm

A 'broad-banded' Lace Monitor.

This familiar monitor is usually grey to dull blue-black with cream or yellow scales scattered across the body. The tail is alternately banded in blue-black and yellow-cream, the bands being narrow in the proximal part and tending to be broader towards the tip. In younger specimens the colour and pattern are much more pronounced. (There is a distinctive colour phase which occurs in northern New South Wales and into Queensland which has broad yellow and black bands along the body and tail.) The lips have a barred appearance. It occurs through the eastern parts of Queensland and New South Wales extending out into the south-western regions. It is also found in south-eastern and northern Victoria and south-eastern South Australia. It is arboreal, found in woodlands and forested areas and is usually seen when foraging on the ground, quickly taking refuge in a tree if alarmed. It shelters in hollows in trees, logs and rock crevices and feeds on nestling birds and birds' eggs, reptiles and small mammals. It also takes carrion and may frequent picnic spots to scavenge for scraps. If provoked it will distend the throat and hiss loudly. Six to 20 eggs are laid in a termite-mound. The female returns to open the nest and release the emerging hatchlings.

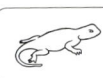

DRAGONS Family AGAMIDAE

These sun-loving diurnal lizards have a skin with a rough dull appearance, often armed with tubercles, prickles or spines. The tongue is broad and fleshy and sometimes sticky to help the lizard catch active insects. Larger species take small lizards, mammals and birds. The tail is usually long and tapered and there are four well-developed limbs. All species are diurnal and lay eggs.

Jacky Lizard (*Amphibolurus muricatus*) 25-35 cm

This species has grey to grey-brown upperparts with a series of lighter patches along the back, sometimes merging to form two vertebral stripes: the tail is often banded. The mouth lining is yellow. Found along the coast and adjacent ranges of southern Queensland, in New South Wales, and through most of Victoria and south-eastern South Australia, it inhabits dry sclerophyll forest, woodlands and coastal heath. It is semi-arboreal and is usually seen on tree branches or stumps. It feeds on insects. Males engage in territorial displays and ritualized combat which consists of head-bobbing, tail-waving and raising the body off the ground. Females deposit up to 12 eggs in a hole dug in sandy soil which is then filled in and concealed.

Nobbi (*Amphibolurus nobbi*) 20-30 cm

This lizard is grey to brown above with a pale cream to yellow stripe along the upper sides. A series of black blotches runs down the back. The inside of the mouth is pink. Breeding males are pink to red behind the hindlimbs, and the side stripe may be a bright yellow. It occurs through north-eastern and south-eastern Queensland, the New England region of New South Wales extending westward to the north-western areas of Victoria and the adjacent areas of South Australia. It inhabits dry sclerophyll forest, coastal sand-dunes and mallee woodlands, and is usually seen among rocks, fallen timber and low vegetation. It feeds on insects. Clutch-sizes range from two to eight eggs.

Mallee Heath Lashtail (*Amphibolurus norrisi*) 25-35 cm

This is grey to grey-brown above, darker on the sides and down the middle of the back. Triangular blotches may extend from the middle of the back out to the sides. The inside of the mouth is yellow. There is a dark line running from the snout, through the eye and ear on to the neck. The tail has distinct bands at least distally. Found in western Victoria and southern regions of South Australia and Western Australia, it occurs in mallee, heath and woodland and is usually found on fallen timber. It shelters in litter, beneath shrubs or timber. It feeds on a wide range of invertebrates and may also eat other lizards. From three to seven eggs are laid in a clutch.

Frilled Lizard (*Chlamydosaurus kingii*) 75-90 cm

Most famous of the dragons, this species is grey to reddish-brown in colour with an obscure pattern or variegation in darker tones. The frill is grey to yellow with vivid orange and red colouration in some populations. Normally the frill is folded back along the body, making the lizard difficult to see when it is lying on a branch. It is erected when the lizard is confronted by a potential aggressor and the act of gaping the mouth (also part of the display) extends the frill. This lizard is also well known for its habit of running on the hindlegs only. The scales are keeled. Found in northern regions of Western Australia and the Northern Territory, and the northern and eastern areas of Queensland as far south as Brisbane, it is an arboreal species which inhabits woodland and dry sclerophyll forest. The Frilled Lizard feeds on insects and small vertebrates. Eight to 14 eggs may be laid per clutch.

Tawny Crevice-dragon (*Ctenophorus decresii*) 15-25 cm

Male

The males and females of this species differ markedly in colour. The male is grey to grey-brown with a black stripe along the side of the body from the neck to the hindlimbs, with cream, yellow or red stripes or blotches above. The lips, chin and throat have a blue, yellow or reddish flush. The female is grey to brown with blackish flecks and has a dark lateral stripe along the body. Occurring on Kangaroo Island, the Mount Lofty Ranges and Flinders Ranges in South Australia and in north-western New South Wales, it favours rocky habitats and feeds on ants, spiders and insects. From three to eight eggs are laid.

Mallee Sand-dragon (*Ctenophorus fordi*) 15-19 cm

This arid-adapted lizard is yellow-brown to red-brown with a pale stripe bordered with black blotches along each side of the back, extending to the base of the tail. Males have black flecks or bars on the throat, and a bar across the chest on to the forelimbs. Occurring in western New South Wales, far north-western Victoria, southern South Australia and south-eastern Western Australia, it is found in sandy, semi-arid areas in association with spinifex in mallee woodland or in chenopod shrubland. It feeds mainly on ants but other invertebrates are also eaten. Most individuals live for only one year but a few survive into the next breeding season. Several clutches of two to three eggs are laid each season.

Military Sand-dragon *(Ctenophorus isolepis)* 15-23 cm

The upperparts of this species are yellow- to reddish-brown, often with brown, black and whitish spots. There is a pale narrow stripe along each side of the back and a mid-lateral stripe which sometimes breaks down into a series of blotches. Males have extensive black colouration on the sides of the head and body. It occurs through western and central Australia to western Queensland, and is found in arid areas, usually in association with spinifex or tussock grasses. It feeds mainly on ants but will take other invertebrates. Females lay one to six eggs and there are at least two clutches in a season.

Central Netted Ground-dragon *(Ctenophorus nuchalis)* 20-28 cm

This dragon is pale red-brown with a darker reticulation over the entire surface. There is usually a pale vertebral stripe. The scales of the lower eyelid are long and spiny, forming a conspicuous fringe. The head and throat of breeding males become orange or red. It occurs extensively through Western Australia, Northern Territory, northern South Australia, western Queensland and western New South Wales in arid regions, usually in open areas on red sandy soils. It is conspicuous because of its habit of perching on top of stumps, rocks or any other elevated site. The diet includes insects, spiders and soft plant material. From two to six eggs are laid per clutch.

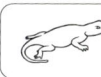

Eastern Two-line Dragon *(Diporiphora australis)* 12-18 cm

The upperparts of this species are grey to reddish-brown with a paler vertebral stripe and a cream to grey stripe along each side of the back. A series of broad brown bars across the body are divided by the three stripes. The throat and belly are whitish, flecked with brown. Occurring in north-eastern Australia from Cape York Peninsula in Queensland to northern New South Wales, it is found in a wide range of habitats from coastal dunes to woodlands. It feeds on insects and spiders.

Canegrass Two-line Dragon *(Diporiphora winneckei)* 15-24 cm

This very slim dragon has a long tail and limbs. It is grey or pale-brown to reddish-brown with a broad grey vertebral band from the nape to the tail. It also has a creamy-yellow stripe along each side of the back, sometimes with dark blotches between this and the vertebral band, and there is often a creamy mid-lateral stripe as well as a pale stripe from the eye to the ear. Occurring in south-western Queensland, northern South Australia, and the southern parts of the Northern Territory to the north-western coast of Western Australia, it is found in arid areas among canegrass or hummock grasses. It feeds on insects. Females lay from two to seven eggs per clutch.

Southern Angle-headed Dragon *(Hypsilurus spinipes)* 25-35 cm

This dragon has a large angular head and a vertebral crest of enlarged spines along the length of the body which is most conspicuous on the nape. Grey-brown to dark-brown in colour, often with a patterning of green or pink, it may also have darker, somewhat obscure transverse bars. It is slow-moving and relies on camouflage to avoid detection. Occurring in the rainforests and adjacent wet sclerophyll forests of south-eastern Queensland and north-eastern New South Wales, it is arboreal and feeds on invertebrates. From two to seven eggs are laid per clutch.

Gilbert's Lashtail *(Lophognathus gilberti)* 35-45 cm

The body of this long-tailed lizard is grey-brown to reddish-brown or almost black with a broad cream or white band along each side of the back extending to the base of the tail: this band may be broken by brown blotches. There is a broad dark-brown bar from the eye to the ear and the jaws are white to cream, often extended as a stripe from the snout to the back of the jaw. There is a vertebral crest from the nape to the base of the tail. Occurring in northern Western Australia, the Northern Territory, Queensland (excepting the east coast), north-western New South Wales and northern South Australia, it is found in woodlands, coastal sand-dunes and stream edges, where it is usually seen perched on branches or other elevated sites. Four to eight eggs are laid in a clutch. (This dragon is sometimes referred to as *Amphibolurus gilberti*.)

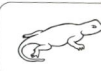

Eastern Water Dragon *(Physignathus lesueurii)* 50-90 cm

This fast-moving dragon has a crest of enlarged spinose scales that extend from the nape to the end of the tail, which is long and laterally compressed. The body colour is grey to olive-green with a series of blackish cross-bars and there is often a distinctive broad black stripe from the eye through the ear to the neck. The throat and belly are often flushed with red or olive-green. Males in southern populations have the throat blotched with orange, red or blue. Occurring in eastern Australia from Cape York Peninsula in Queensland to eastern Victoria, it is diurnal and semi-aquatic, being found in the vicinity of watercourses and along the coast in some areas. It feeds on insects, fruit, crustaceans, fishes and small reptiles and forages even in rock-pools on the seashore. Two or more clutches of six to 20 eggs may be laid in a season.

Left: *Male*. Above: *Hatchlings*.

Western Bearded Dragon *(Pogona minor)* 30-40 cm

This bearded dragon has a poorly developed 'beard' and is grey to greyish-brown in colour. Juveniles have a more prominent pattern with blotches on either side of the vertebral area. There is a dark stripe from the eye to the neck and the tail is obscurely banded with pale grey. Occurring in central Western Australia through to western Northern Territory and South Australia, it inhabits open forest, woodland and shrublands and feeds on arthropods and plant material.

Eastern Bearded Dragon *(Pogona barbata)* 35-55 cm

This dragon has a large triangular head and a well developed 'beard' with numerous rows of enlarged spines across the throat. The spines across the back of the head are in a backward-curving arc. It is grey, yellowish-brown or red-brown in colour. Juveniles have a pattern consisting of a series of pale blotches from the nape to the tail, but this pattern is often absent or obscure in adults. The belly is grey with scattered darker circles. The inside of the mouth is yellow. Occurring in the eastern areas of Queensland and New South Wales, north-eastern Victoria and south-eastern South Australia, it is found in woodland and open forest. Semi-arboreal in habit, it is often seen on stumps or fallen timber. It feeds on insects, small reptiles and flowers. From eight to 35 eggs are laid in a burrow which is then sealed by the female.

Central Bearded Dragon *(Pogona vitticeps)* 40-55 cm

This large, robust dragon has a grey to orange-brown colouration. Juveniles are patterned with blotches along the vertebral region which may merge to form stripes, a patterning which is obscure or absent in adults. Some animals have orange colouring on the side of the head, with a distinct bright orange ring around the eye. The spines across the back of the head are in a straight line. Occurring through central New South Wales, central Queensland, south-eastern areas of the Northern Territory, northern Victoria and eastern South Australia, it is found in a wide range of semi-arid to arid habitats including woodlands, mallee and hummock grasslands. It often perches prominently on fence-posts or stumps. Females lay up to two clutches of 11 to 25 eggs in burrows.

Thorny Devil *(Moloch horridus)* 15-19 cm

This bizarre lizard cannot be confused with any other. There are large thorn-like spines over the head, body, limbs and tail, and a prominent hump on the nape. It is grey-brown to reddish-brown with a pale vertebral stripe and has a prominent pattern with dark-edged brown blotches. Occurring through most of Western Australia, southern areas of the Northern Territory, western Queensland and western and northern South Australia, it is a slow-moving lizard found in open woodland, shrubland and hummock grasslands. It feeds exclusively on small black ants. Females lay three to ten eggs in specially excavated burrows.

Blotch-tailed Earless Dragon *(Tympanocryptis cephalus)* 10-14 cm

This is a small robust dragon with a short tail which tapers rather abruptly at the base. It is greyish-brown to bright reddish-brown with a broad dark-brown band across the neck. Conical tubercles on the back are arranged more or less in rows. The tail has a series of dark crossbands. Occurring in central Western Australia, southern Northern Territory, western Queensland and northern South Australia, it inhabits arid to semi-arid stony areas. It is known to feed on grasshoppers.

Mountain Heath-dragon *(Tympanocryptis diemensis)* 15-20 cm

With a short rounded head and a blunt snout, this dragon is light grey to red-brown with two broad, lighter-coloured stripes along each side of the back. These stripes may be deeply scalloped along the edge facing the midline of the back. The mouth lining is blue and the tongue orange in many populations. It occurs in the highlands of southeastern Australia from northern New South Wales through to Victoria, as well as in eastern Tasmania. It is found in cool well-timbered habitats, often in areas that receive winter snow. It feeds on ants and other arthropods. Females lay from four to six eggs per clutch in late spring.

Lined Earless Dragon *(Tympanocryptis lineata)* 8-16 cm

Light grey to dark brown in colour with darker cross-bars on the body and tail, this dragon has up to five white stripes along the length of the body although some of these may be broken or obscure. The scales are keeled and there are numerous spiny tubercles on the body. Some individuals may have yellow colouration on the lips and throat. Occurring in South Australia, eastern Western Australia, the Northern Territory, western Queensland, northern Victoria and the western and southern areas of New South Wales, it inhabits a wide variety of habitats including open woodland, grasslands and spinifex. It perches on stones and timber and shelters in earth cracks and in ground litter and feeds on arthropods. Females lay clutches of nine to 11 eggs.

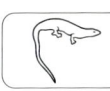

SKINKS Family SCINCIDAE

The skinks are the largest and most diverse family of lizards in Australia and it is difficult to generalize on their features. There are diurnal, nocturnal and crepuscular species, and they may be arboreal, terrestrial or fossorial. In size they range from a few centimetres to more than 50 cm. Most have smooth scales and a long tail that is easily cast off; some species have rough or very prickly scales. A notable area of diversity is in the number and size of the limbs (four, two or none) and the number of digits on each limb.

Three-clawed Worm-skink (*Anomalopus verreauxii*) 20-38 cm

This is pink to brown above with a yellow or cream band on the nape (which may be obscure in older animals). The throat, chin and underside of the body are often yellow. The limbs are short, the forefoot having three digits, and the hindfoot one. The tail is readily discarded if the animal is handled roughly or picked up by the tail. Occurring on the coast and adjacent ranges of north-eastern New South Wales and south-eastern Queensland, it inhabits sclerophyll forest, woodland and the edges of rainforest. A burrowing lizard, it prefers soft soils, particularly those with a high humus content, beneath logs or rocks. It feeds upon invertebrates found in the the soil. The clutch is three to 11 eggs.

113

Lined Rainbow-skink *(Carlia jarnoldae)* 10-13 cm

Mature males of this species are brown above with five to seven narrow black stripes extending from the neck to the tail. These stripes may sometimes be broken. The upper sides are black with scattered blue spots and the lower sides are orange. The lips and throat are a pale green. Females and non-breeding males are dark brown above, the head being a lighter coppery-brown. There are scattered lighter and darker flecks along the back. The upper sides are black and there is a prominent white stripe along the middle of the sides. There are four digits on the forefoot and five on the hindfoot. Occurring in north-eastern Queensland from the Townsville area to Cape York Peninsula, it inhabits dry sclerophyll forest and woodland, sheltering in leaf litter and under rocks. It is often seen foraging among leaf litter or around logs and fallen trees.

Shaded-litter Rainbow-skink *(Carlia munda)* 6-10 cm

This is grey to dark brown with black, pale-brown and white spots and flecking. A thin white stripe runs from the lips to at least the forelimb. In breeding males the throat and chest may be bluish with black margins to the scales. The sides of breeding males are red from the forelimb to the hindlimb; females and non-breeding males do not have this colouration. There are four digits on the forefoot and five on the hindfoot. Occurring through the northern regions of Western Australia and the Northern Territory, and the northern and eastern areas of Queensland, it is found in areas of dry sclerophyll forest, woodland and hummock grass. It is usually seen foraging in ground litter or on grass tussocks, and it feeds on insects and spiders. The female lays two eggs per clutch.

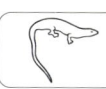

Southern Rainbow-skink *(Carlia tetradactyla)* 10-15 cm

Deep brown in overall colour, this skink has a paler stripe along the upper edge of each flank. The middle of the back between these stripes is marked with three to four rows of white spots or dashes, edged in black. There is a broad orange stripe on the upper and lower sides and in breeding males the lower sides are a blue-green colour. The forefoot has four digits, the hindfoot five. Occurring in northern Victoria, through New South Wales to south-eastern Queensland, it inhabits dry sclerophyll forest and woodland with an understorey of tussock. It is usually found beneath timber or rocks or foraging in the litter. It feeds on insects and other invertebrates. The usual clutch is two eggs.

Callose-palmed Shinning-skink *(Cryptoblepharus plagiocephalus)*
10-12 cm

This small sun-loving lizard has a flattened body. Brown or grey above, it usually has a narrow, ragged, pale stripe along each side of its back and numerous scattered pale and dark flecks over the body. The lower surfaces of the fore- and hindfeet are brown. All feet have five digits. It occurs in Western Australia, western areas of South Australia, the Northern Territory except for the south-eastern regions, and northern Queensland. It is found in tropical woodland, eucalypt forest and often on man-made structures. It is arboreal or rock-inhabiting and preys upon arthropods. The female lays two eggs per clutch.

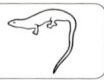

Cream-striped Shinning-skink *(Cryptoblepharus virgatus)* 8-10 cm

A markedly flattened body and a distinct cream or white stripe (bordered dorsally by a broad black stripe) along each side of the back distinguishes this lizard. The base colour ranges from silver-grey to brown to almost black. The lower surfaces of the fore- and hindfeet are whitish and there are five digits on each foot. Occurring through eastern and southern Australia from Cape York Peninsula in Queensland to southern Western Australia, it is found in wooded areas and rock outcrops but most often seen on buildings, fences and walls. There are two to three eggs in a clutch.

Southern Mallee Ctenotus *(Ctenotus atlas)* 12-18 cm

This skink is dark brown to black above with eight or ten white stripes on the back and sides. The head, tail and limbs are brown. Each foot has five digits. Occurring in the central and south-western areas of New South Wales, western Victoria, southern South Australia and the southern areas and interior of Western Australia, it is found in mallee woodland with spinifex understorey. It shelters within spinifex clumps and is known to construct shallow burrows beneath the clumps. It forages for invertebrates in the early part of the day. There are one to two eggs in a clutch.

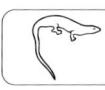

Leopard Ctenotus *(Ctenotus pantherinus)* 20-30 cm

This skink is grey to brown in colour with a series of black-edged white spots which also occur on the sides. In some populations these may be replaced by rows of black lines or dashes. There are five digits on each limb. Occurring through most of Western Australia (apart from the southern regions and the far north), the Northern Territory (except for the far north), northern areas of South Australia, central and western areas of Queensland and far north-western New South Wales, it inhabits areas with a ground cover of spinifex in which it normally shelters. It feeds on invertebrates, termites making up a large part of the diet. The clutch is three to nine eggs.

Robust Ctenotus *(Ctenotus robustus)* 24-35 cm

This active diurnal skink has a conspicuous broad, black vertebral stripe from the nape to the tail: this is bordered by a narrow white to cream stripe. The basic colour ranges from fawn to brown and there is a pale stripe along each side of the back from above the eye to the tail. The upper parts of the sides are dark brown with a series of pale blotches, and the lower flanks are creamy brown. Occurring through south-eastern South Australia, Victoria, eastern New South Wales and Queensland, the northern areas of the Northern Territory and north-western Western Australia, it is found in open woodland, forest, grassland and coastal dunes. It shelters in short burrows under rocks or timber. It feeds on insects and lays four to seven eggs in a clutch.

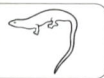

Barred Wedge-snout Ctenotus *(Ctenotus schomburgkii)* 12-15 cm

This skink has a prominent white stripe along each side of the back from above the eye on to the tail and another stripe along the middle of each flank from above the eye to the tail. The area in between is black with a single row of large orange-brown blotches. Typically it is red-brown above with five black dorsal stripes from the nape to the tail. (In some populations these stripes may be absent or broken.) Occurring in southern Western Australia, southern Northern Territory, South Australia, south-western Queensland, western New South Wales and north-western Victoria, it is found in shrubland, mallee and spinifex or tussock grassland. It feeds mainly on termites and lays two to four eggs in a clutch.

Coppertail *(Ctenotus taeniolatus)* 10-19 cm

The pattern of this skink consists entirely of longitudinal stripes, without spots or blotches, and the tail is typically orange or reddish in colour (brown or olive in some populations). It is brown above with a black vertebral stripe edged with a narrow cream stripe. There is a cream stripe along each side of the back, edged in black above, and two white stripes with black in between along the middle and lower middle of each side. Occurring in eastern Queensland from Cape York Peninsula south to eastern New South Wales and northern Victoria, it is usually associated with rocky areas where there are surface rocks and sandy soil in open forest, woodland and heathland. It is found in short burrows under rocks. It feeds on arthropods and lays two to seven eggs in a clutch.

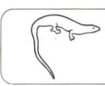

Common Slender Bluetongue (*Cyclodomorphus branchialis*) 20-30 cm

Slender and short-limbed, this skink is pale grey to olive-brown above, paler on the sides. The edge of each scale is usually dark or there may be a dark spot on each scale. The underparts are white, yellow or orange-brown.

Young specimens often have numerous whitish spots on the body and tail. Occurring through Western Australia, South Australia, most of the Northern Territory, western Queensland and far-western New South Wales, it is found in woodland, heath, coastal sand-dunes and

spinifex grassland, where it shelters in spinifex hummocks or leaf litter. It feeds on a wide range of invertebrates and also takes small lizards. Females produce from two to five young.

She-oak Skink (*Cyclodomorphus casuarinae*) 30-40 cm

This slender skink has short limbs and a long tail. The basic colour can be grey through reddish-brown to almost black. The edges of the scales are usually dark-edged, producing a series of narrow longitudinal dark lines. The belly may have an orange or yellow tinge. The tip of the snout is often darker and there may be a dark bar beneath the eye. Juveniles have a black band on the neck. Occurring in northern Tasmania, eastern Victoria and eastern New South Wales, it is found in dry sclerophyll forest, heathland and coastal sand-dunes. It is sometimes found in high densities on grazing lands. For most of the day it shelters under ground debris or within a

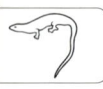

A female She-oak Skink with a litter of young.

dense mat of tussock grasses. Crepuscular to nocturnal, it feeds upon insects and possibly small lizards. It is snake-like in appearance when it moves through thick grass with its legs folded back against the body. When threatened it mimics the defensive posture of a juvenile Eastern Brown Snake *(Pseudonaja textilis* – see page 53). Females produce from four to seven (occasionally up to 19) young.

Major Skink *(Egernia frerei)* 30-40 cm

This large skink is light to dark brown. The scales on the back usually have a darker streak, forming narrow longitudinal lines from the neck to the tail. The upper lateral area is dark brown becoming paler on the lower flanks. (Some populations have numerous whitish spots on the sides.) The lips are white or cream, barred with brown, and the ear lobules are cream. Occurring in north-eastern New South Wales, eastern Queensland to Cape York Peninsula, and the Torres Strait islands, it is also found in Arnhem Land in the Northern Territory. A wary lizard which is not often seen, it is found in woodlands and vine thickets where there is a dense ground cover and fallen timber or rock. It feeds on insects and vegetable matter. Up to three young are produced by females in summer.

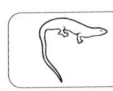

Cunningham's Skink *(Egernia cunninghami)* 30-40 cm

A large robust skink whose tail is about the same length as its body. The scales on the back, sides and tail are keeled with a single sharp spine which is more pronounced on the tail. It is extremely variable in colour, ranging from light brown to black. In some populations there is no pattern while others have a pattern consisting of dark-brown scales and cream spots. The darker scales may form irregular cross-bands across the body. In other populations the cream spots may form cross-bands. The eyelids are edged in white and the belly is pink to orange. Occurring on the slopes and mountain ranges of New South Wales, western Victoria, south-eastern Queensland and the Mt. Lofty ranges and Fleurieu Peninsula in South Australia, it is usually found in rocky areas, where it inhabits crevices and exfoliations. In some areas it utilizes hollows and cracks in dead trees. Its habitat is woodland, heathland and open forest, and it is diurnal and gregarious. It feeds on insects and vegetable matter. Females give birth to from two to six young.

Head of E. cunninghami *showing the white-edged eyelids.*

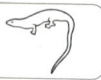

Unadorned Desert-skink *(Egernia inornata)* 15-20 cm

This interesting desert lizard has a blunt head and robust body and is fawn to red-brown in colour. Dark edges of the scales form narrow longitudinal lines. Scattered white and black flecks may be present. The sides are brown with black and white spots and the black spots may form indistinct cross-bars. The lips are light and are barred with dark brown. Occurring in southern Western Australia, southern Northern Territory, northern South Australia, western Queensland, western New South Wales and north-western Victoria, it is found in shrubland, mallee woodland and hummock grassland in arid and semi-arid areas. It makes a series of burrows at the base of a spinifex clump or a shrub, from which it forages. It feeds on arthropods and females produce from two to four young.

King's Skink *(Egernia kingii)* 30-50 cm

This very large brown to almost black skink has distinctively keeled scales. Individuals may lack dorsal markings entirely or have a series of fine indistinct lines, or prominent creamy yellow spots and streaks. It is cream to grey below with darker streaks, particularly on the throat. Occurring in south-western Western Australia and nearby offshore islands, it inhabits granite outcrops and heathlands and is found in rock crevices or burrows. It feeds on insects, vegetable matter, other lizards and the eggs of seabirds. Females give birth to from two to eight young.

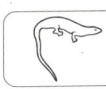

Land Mullet *(Egernia major)* 40-60 cm

This fascinating lizard, one of the largest of the skinks, acquired its strange name because of its highly polished body and slithery locomotion. It is shiny black or dark brown on the back and sides and the eye has a white to cream rim. Young animals have white spots on the sides, but these disappear as they mature. Occurring in south-eastern Queensland and northern coastal New South Wales, it is found in rainforest or adjacent wet sclerophyll forest. It is a shy and secretive lizard which may be seen basking in the sun, usually near dense vegetation. It shelters in burrows or logs and feeds on insects, snails, fruit and fungi. From two to nine young are produced by the female.

South-western Crevice-skink *(Egernia napoleonis)* 25-35 cm

The upper body scales of this skink are strongly keeled. It is olive-brown to dark grey-brown in basic colour with black spots forming three lengthwise rows on the back. There are scattered white spots on the sides and sometimes on the back. A black stripe from the snout passes below the eye and broadens along the upper flanks above the forelimbs. Occurring in south-western Western Australia in the coastal and adjacent regions, it inhabits woodland, forest, heathland and rock outcrops. It is often arboreal or rock-dwelling. It feeds on invertebrates, vegetable matter and smaller lizards. The females give birth to from two to four young.

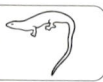

Gidgee Skink *(Egernia stokesii)* 20-27 cm

This robust skink has a strongly depressed tail which is considerably shorter than the body. The dorsal scales are keeled with a short spine which is more developed on the tail scales. It is olive-brown to red-brown or black above, often with scattered paler and darker scales forming irregular cross-bands. It occurs in widely separated populations in south-western Queensland, north-western New South Wales, northern and eastern South Australia, southern Northern Territory and the mid-western coast of Western Australia and into the interior. Inhabiting rocky outcrops and stony hills in deep crevices, in hollow trees and splits, it is diurnal and usually observed basking near its shelter site. It can be very difficult to dislodge when in a crevice because of its spiny scales. It feeds on insects and vegetable matter and the females produce between one and three young.

Tree Crevice-skink *(Egernia striolata)* 15-24 cm

This lizard has a somewhat flattened head, body and tail: the scales are weakly keeled and are not glossy. The upperparts are basically dark brown to grey, with scattered pale flecks. There is a broad lighter area at the sides of the back from the nape to at least mid-body and a dark upper lateral band from the eye to the hindlimb. The lips are white, edged in brown. The belly is yellow to orange. Occurring through most of New South Wales (except for the coast and high ranges), northern Victoria, eastern South Australia, and eastern and central Queensland, it prefers woodland, dry sclerophyll forest and rock outcrops under loose bark or rock, in tree hollows and in crevices. It is often found in groups comprising both adults and juveniles. It feeds on invertebrates and probably other lizards. The females produce from two to six young.

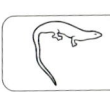

White's Skink *(Egernia whitii)* 20-30 cm

The eyes of this skink are prominently ringed with cream and there are scattered black-edged white spots along the grey flanks. It is grey-brown to red-brown or almost black above either with no markings other than a few black flecks, or with two wide dark stripes on either side of the midline of the back, each stripe enclosing a series of creamish spots. There is a pale streak along the upper lips to the ear. Occurring in south-eastern Queensland, eastern New South Wales, Victoria, Tasmania and the Bass Strait islands, this skink favours woodland, dry sclerophyll forest, heathland and tussock grassland, particularly where rock is present. It lives in a burrow system beneath rocks or in stumps or logs and several animals may utilize the same burrow. It eats invertebrates and vegetable matter. The females produce from two to five young.

Broad-banded Sand-swimmer *(Eremiascincus richardsonii)* 15-27 cm

This 'sand-swimming' lizard is usually located in loose sand in creek beds, animal burrows or beneath shrubs. Its name arises from its usual escape method of burrowing into loose sand in a wriggling motion. It has shiny scales and short, well-developed limbs. Basically pale brown to red-brown, it has eight to 14 dark cross-bands which may be complete or broken. There are 19 to 32 similar bands on the tail (if it is original). Occurring through most of Western Australia, Northern Territory and South Australia, south-eastern Queensland, western New South Wales and the far north-west of Victoria, it inhabits woodland, shrubland or hummock grassland on harder, rocky ground. It feeds on invertebrates and other lizards, lying in wait with only its head above the sand, then erupting on to passing prey.

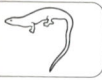

Warm-temperate Water-skink *(Eulamprus heatwolei)* 10-20 cm

This skink is coppery-brown above with numerous flecks of black on the head and scattered black spots on the body and tail. A pale stripe extends from above the eye to above the fore-

limbs. The upper lateral area is black with cream spots. The lower lateral area is cream with black flecks. The belly is white with blackish blotches on the throat and chin. The pelvic region and thighs may be bright yellow in adults. It occurs in central and eastern Victoria and southern New South Wales to the Blue Mountains, with an isolated population in the high country in northern New South Wales and another in south-eastern South Australia. Found along creeks or seepages and swamps in a wide range of habitats including woodland, wet and dry sclerophyll forest, heathland and tussock grassland, it is usually seen basking or foraging. It can occur in high densities in some areas. It feeds mainly on invertebrates but occasionally on vertebrates such as small skinks, small fish and tadpoles. Females produce two to six young.

Alpine Meadow-skink *(Eulamprus kosciuskoi)* 10-18 cm

This skink has a blunt rounded head and is olive-brown with a narrow black vertebral stripe. A pale stripe along each side of the back is edged above with black. The upper sides are black with whitish

flecks: the lower sides are cream to yellowish with some black scales. The underparts are cream to yellow with grey mottling on the throat and black flecks on the chest and belly. In the northern form, the vertebral stripe is missing or only obscurely present. Occurring in the Snowy Mountains region on the Victorian/New South Wales border and the New England plateau and Barrington Tops region of New South Wales, it inhabits woodland, heathland and tussock grassland, usually near bogs or along the edge of streams and around logs in woodlands. It feeds on invertebrates and females produce two to five young.

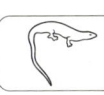

Eastern Water-skink (*Eulamprus quoyii*) 20-30 cm

The upperparts of this skink are golden olive-brown, usually with scattered dark flecks. A narrow whitish to pale yellow stripe extends from the eye along each side of the back to the tail, but may fade out midway along the body. The upper lateral area is black with a series of cream or yellow spots. The underparts are white to yellow, often with grey or black flecks. It occurs in eastern Australia from northern Queensland to southern New South Wales, extending through the Murray-Darling river-system to north-western Victoria and south-eastern South Australia. Found in a wide range of habitats, it is most common along the margins of watercourses or swamps and may also be seen foraging down to the intertidal zone in some localities along the coast. It hides in crevices, burrows and among rocks and timber. It feeds on arthropods, tadpoles and berries and the females produce two to nine young.

Bar-sided Forest-skink (*Eulamprus tenuis*) 10-18 cm

This skink is fawn-grey to deep-brown above with scattered dark blotches forming a variegated pattern. The dark brown of the upper lateral area is broken into large blotches. The lower lateral surface is creamy grey with darker marbling. The underside is cream to yellow. The lips are pale with dark bars. It ranges along the east coast from Cape York Peninsula in Queensland to southern New South Wales and inhabits wet sclerophyll forest, rainforest and the moister areas of woodlands and dry sclerophyll forests. It can be found in rock crevices and exfoliations, often exploiting man-made structures. Some populations utilize hollows, cracks or loose bark in trees. It usually forages early in the morning or late afternoon. It feeds on arthropods and the females produce two to six young. (This skink is sometimes referred to as *Sphenomorphus tenuis*.)

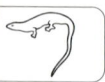

Northern Bar-lipped Skink (*Glaphyromorphus isolepis*) 10-20 cm

P. Griffin

The back and flanks of this skink are tan to orange-brown, with darker-brown flecks. Usually there is an obscure dark zone along the sides of the back. The lips are barred with cream and brown and the ear opening is conspicuous. Distributed through the northern parts of the Northern Territory and north-western Western Australia, it inhabits woodlands and shrublands. It is often found near the edges of swamps and watercourses under leaf litter and other ground debris. It is nocturnal and feeds on arthropods. Females lay two to seven eggs in a clutch. (This skink is sometimes referred to as *Sphenomorphus isolepis*.)

Prickly Forest-skink (*Gnypetoscincus queenslandiae*) 10-18 cm

G. A. Hoye

The scales on this species are keeled with a point, giving rise to the common name. It is dark brown to blackish, normally with irregular pale cross-bands. The lips and throat are dark brown with some pale spots and flecks. It is restricted to north-eastern Queensland in highland and lowland rainforests. A very secretive lizard, it avoids direct sunlight and is found in damp, shaded conditions under rotting logs, rocks or in leaf litter, where it feeds on slugs, snails and arthropods. Females produce one to five young. (This skink is sometimes referred to as *Tropidophorus queenslandiae*.)

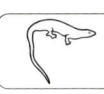

Three-toed Earless Skink (*Hemiergis decresiensis*) 6-11 cm

This slender, short-limbed lizard has three digits on each limb. It is generally grey-brown to chocolate-brown, with a series of darker longitudinal lines. A black stripe extends from the eye along each side of the back to the tail tip. The sides are grey with numerous darker flecks and the belly is cream to yellow or orange. The scales on the throat are dark-edged and the undersurface of the tail is spotted with dark brown. Occurring in the highlands and tablelands of New South Wales,

central Victoria and south-eastern South Australia, it inhabits wet and dry sclerophyll forest and woodland. It is a burrowing species and may normally be found under stones or timber. It feeds on arthropods and the females produce one to five young.

Lowlands Earless Skink (*Hemiergis peronii*) 10-20 cm

This skink has four digits on each limb (except for the populations in far-western Western Australia, which have three). It is brown to olive-brown above with a series of longitudinal lines made up of darker spots. There is a narrow black line from the eye along the sides of the back to the tail: this may be tinged with yellow to orange above. The flanks are pale grey-brown with scattered black flecks and the belly is yellow, each scale having a dark-brown edging which contributes to a conspicuous pat-

tern. Occurring in south-western Victoria, southern South Australia and southern Western Australia, it inhabits woodland, heathland and coastal dunes where it is found in sandy soil under rocks or timber. It feeds on arthropods and the females produce one to five young.

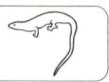

Pink-tongued Skink (*Hemisphaeriodon gerrardii*) 30-45 cm

G. E. Schmida

This skink has a broad angular head, short limbs, a slender body and a long, prehensile tail. It is born with a bright blue tongue and although this usually changes to pink between one and two years of age, occasional older animals may be found still with a blue tongue. Adult colouration is variable, ranging from silver-grey to light brown, with cross-bands of black or brown which are indistinct and sometimes only on the sides: some individuals have no bands at all. In contrast, juveniles are distinctively marked, being a very pale brown with 17 to 24 sharp-edged dark cross-bands from the snout to the tip of the tail. Occurring along the east coast from the Blue Mountains west of Sydney in New South Wales to the Cairns district in Queensland, it is found in wet sclerophyll forest and rainforest and also in woodland where suitably moist and humid conditions occur. It shelters in tree hollows, hollow logs and crevices, among rock outcrops, and in deep leaf litter. Terrestrial and partially arboreal, it forages at dusk and during the first few hours of darkness (often basking during the early part of the day and late in the afternoon). It feeds predominantly on slugs and snails. Females produce litters of four to 67 young.

Dark-flecked Garden Sunskink (*Lampropholis delicata*) 6-10 cm

Dark grey-brown to deep brown above, this skink usually lacks markings but occasionally has a narrow, darker vertebral stripe and there may be a scattering of pale and dark flecks. The upper lateral area is dark brown, often with a pale stripe along each side of the back. A pale stripe may also be present along each side. The belly is whitish with brown flecks. Occurring through eastern and north-eastern Queensland, eastern New South Wales, southern Victoria, eastern Tasmania and south-eastern South Australia, it is found in wet and dry sclerophyll forest, woodland and heathland. It is also a very common inhabitant of suburban gardens. It inhabits leaf litter and other ground debris and feeds on arthropods. From two to six eggs are laid, often in a communal nest.

Pale-flecked Garden Sunskink (*Lampropholis guichenoti*) 6-9 cm

The upperparts of this skink are grey-brown to coppery-brown, usually with a dark vertebral stripe. There is normally a scattering of paler and darker scales. A dark-brown stripe runs from the nostril through the eye and along the flanks to the base of the tail: this is bordered above and below by a pale stripe. The head is often a brighter coppery-brown. Occurring in south-eastern Queensland, eastern New South Wales, Victoria and south-eastern South Australia, it is found in dry sclerophyll forest, woodland, heathland and tussock grassland. It is also a very common inhabitant of suburban gardens. It feeds on arthropods and lays two to six eggs, sometimes in a communal nest.

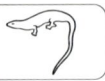

South-eastern Slider *(Lerista bougainvillii)* 10-12.5 cm

This slim, elongated lizard has short limbs, each with five digits. It is basically pale silver-grey to pale grey-brown, either with no pattern or with a series of longitudinal lines made up of dark-brown dots. There is a broad black band from the nostril through the eye and along the flanks to the base of the tail. In some populations the tail is yellow- or orange-brown with a freckled appearance; in others it is the same colour as the body. Occurring through south-eastern New South Wales, Victoria, north-eastern Tasmania and the Bass Strait islands, and south-eastern South Australia (including Kangaroo Island), it is a burrower found in dry sclerophyll forest and woodland in loose soil under rocks, logs and leaf litter. It feeds on arthropods. It is interesting to note that while the females of the Tasmanian and Kangaroo Island populations give birth to two to four young, females on the mainland lay eggs (also two to four to a clutch).

North-western Sandslider *(Lerista bipes)* 10-13.5 cm

This skink has no forelimbs, and the short hindlimbs have only two digits. It is fawn to pale reddish-brown in colour, usually with two lines from the nape to the tail made up of dark-brown dots. There is a dark-brown lateral stripe from the snout through the eye to the tail. The tail is flushed with grey to yellow and the snout protrudes beyond the lower jaw. Occurring through central and northern Western Australia, Northern Territory, far-western Queensland and north-western South Australia, it is found in hummock grasslands on sand-dunes and plains where it shelters beneath leaf litter and ground debris or grass hummocks. The usual clutch is two eggs.

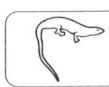

Wood Mulch-slider (*Lerista muelleri*) 9-11.5 cm

The short fore- and hindlimbs of this interesting lizard each have three digits. It is grey to olive-brown or black, often with four dark lines or rows of spots. There is a blackish-brown stripe along the upper flanks and below this the lower flanks are paler than the body colour. Occurring in south-western Queensland, western New South Wales, north-western Victoria, South Australia, the southern areas of the Northern Territory, and central Western Australia through to the coast, it is found in open woodlands, shrublands and grasslands in drier areas, where it burrows in loose soil and debris under and in rotting timber. It feeds on arthropods and lays one or two eggs in a clutch.

Keeled Slider (*Lerista planiventralis*) 11-15 cm

The short forelimbs of this skink have only two digits while the hindlimbs have three. The snout is depressed and protrudes beyond the lower jaw and there is a distinctive keel on either side of the flattened belly. Grey to brown in basic colour this slider has four narrow dark-brown to black stripes from the nape to the tail. The sides are fawn with a blackish stripe from the snout through the eye to the tail. Occurring in the coastal regions of mid-Western Australia, it is found in coastal dunes and low shrubland. It shelters beneath stumps and in sand beneath leaf litter. The clutch is two to four eggs.

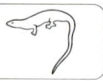

Common Dwarf Skink (*Menetia greyii*) 5-10 cm

This skink has four digits on the forefoot and five on the hindfoot. It is metallic-brown, often with a number of darker longitudinal stripes made up of spots. There is a blackish stripe along the upper flanks from the nostril to the tail, bordered below by a white stripe at least to the forelimb. Breeding males have an orange throat and chest which is otherwise cream to yellow. Occurring in Western Australia, Northern Territory, South Australia, northern Victoria, New South Wales west of the ranges, and most of Queensland, it inhabits dry sclerophyll forest, woodland, shrubland and grassland. It is usually seen in leaf litter and other ground cover. The clutch is one to four eggs.

South-eastern Morethia (*Morethia boulengeri*) 9-12 cm

The back of this skink is grey to brown, often with some scattered paler and darker flecks. The upper lateral area is blackish and there is a distinct white stripe along the middle of each side with a narrow black stripe below. The tail is often reddish-brown and in breeding males the throat and chin are a bright red-orange. Occurring in eastern Western Australia, southern Northern Territory, South Australia, northern Victoria, New South Wales west of the ranges and south-western Queensland, it is found in dry sclerophyll forest, woodland and shrubland where it is usually seen in leaf litter and ground debris. It feeds on arthropods and lays one to six eggs in a clutch.

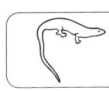

Lined Firetail-skink (*Morethia ruficauda*) 6-9 cm

The tail, lower back and hindlimbs of this skink are a distinctive orange-brown to bright red. The dorsal surface of the body is a glossy black sometimes with a pale vertebral stripe. There are two wide yellowish-white stripes along the sides with a black band between them. Occurring in northern Western Australia, western Northern Territory and north-western South Australia, it is found in woodland, shrubland and spinifex in the vicinity of rocky outcrops and hills. It feeds on arthropods and females lay one to three eggs per clutch.

Red-throated Cool-skink (*Bassiana platynotum*) 14-20 cm

The red or orange throat and chest is usually quite conspicuous in this lizard. It is basically light grey to brown in colour and the outer edges of the scales are sometimes black, creating a reticulated pattern. A broad black stripe extends along the upper flanks from the nose through the eye to the tail. The lower sides are grey and the belly is white. Occurring in eastern Victoria, eastern New South Wales and extending slightly into Queensland, it is found in dry sclerophyll forest, woodland and grassland, usually in leaf litter or other ground debris or under rock slabs. It feeds on invertebrates and lays three to ten eggs in a clutch. (This lizard is sometimes referred to as *Leiolopisma platynotum*.)

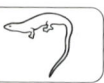

Tussock Cool-skink *(Pseudemoia entrecasteauxii)* 12-17 cm

This skink is light brown to olive-brown with a darker vertebral stripe and a pale stripe along the upper parts of the sides. The upper flanks are brown with a whitish mid-side stripe from the ear to the tail. The lips are whitish. The belly is light grey, often with an orange-red flush. In mature males the mid-side stripe is flushed with red or orange, at least towards the front of the body. Some populations are a darker brown to almost black, the pattern being obscure or almost absent. It occurs in the highlands and southern areas of New South Wales, southern Victoria, Tasmania, the Bass Strait islands, and south-eastern South Australia. Found in dry sclerophyll forest, woodlands, heathlands and alpine grasslands, it shelters in leaf litter and tussocks and among fallen timber. It feeds on insects and females produce one to seven young in a litter. (This lizard is sometimes referred to as *Leiolopisma entrecasteauxii*.)

Yellow-bellied Three-toed Skink *(Saiphos equalis)* 15-20 cm

This smooth and shiny skink has very short limbs, each with three digits. It is grey-brown to chocolate-brown, often with small darker spots which form longitudinal lines. The sides are a darker brown to black and the belly is yellow to orange. The tip of the tail is black. It occurs in eastern New South Wales and south-eastern Queensland. A burrowing lizard, it is found in wet and dry sclerophyll forest and heathland, usually under logs, rocks etc. It feeds on invertebrates and worms. Females lay one to seven large eggs which hatch two to eight days after being laid.

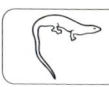

Weasel Skink (*Saproscincus mustelina*) 11-16 cm

This lizard has a distinctive creamy-white streak at the back corner of each eye and usually an orange to red stripe along the upper parts of each side above each hindlimb, joining on the tail. It is russet-brown above, usually with scattered pale flecks. The belly is white to yellow. Occurring in eastern New South Wales and southern and eastern Victoria, it is found in wet and dry sclerophyll forest and is common in some suburban gardens. It prefers moister, cooler areas and is secretive, usually only active in the early evening or on overcast days. It shelters in leaf litter and ground debris, preferring areas where there is quite dense ground cover. It feeds on arthropods and the females lay two to seven eggs in a clutch, sometimes communally. (This lizard is sometimes referred to as *Lampropholis mustelina*.)

Centralian Bluetongue (*Tiliqua multifasciata*) 35-45 cm

In common with other members of the blue-tongue genus this skink has a broad triangular head, a stout body with relatively short limbs, a short tail and a blue tongue. It is grey to pale brown with a series of orange-brown cross-bands and a broad wedge-shaped black streak from the eye to above the ear. Occurring in the Northern Territory, northern Western Australia, northern South Australia, western Queensland and the far north-western region of New South Wales, it is found in open woodland, shrubland and hummock grassland in arid and semi-arid habitats. It shelters in animal burrows, leaf litter and grass hummocks and feeds on insects, carrion and vegetable matter. There are two to ten young in a litter.

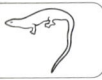

Blotched Bluetongue *(Tiliqua nigrolutea)* 35-50 cm

The Blotched Bluetongue is typical of this genus, with its broad triangular head, stout body, relatively short limbs, short tail and blue tongue. It is dark brown to black with a series of large yellow to pink blotches on the back extending on to the tail (usually as bands). The belly is cream with black mottling. The head which is paler than the body, has numerous black spots: the lips are cream to pale yellow. It occurs in south-eastern South Australia, southern Victoria, north-eastern Tasmania, the Bass Strait islands, and the highlands of southern and central New South Wales. Found in dry sclerophyll forest, woodland, grassland and heathland, it shelters in hollow logs, burrows, under rock and in leaf litter. This is the most cold-adapted of the bluetongues and it is active at low temperatures. It feeds on insects, snails, carrion, fungi and vegetable matter. Females produce five to 12 young in a litter.

Common Bluetongue *(Tiliqua scincoides)* 45-55 cm

Largest of the bluetongues and one of the world's largest skinks, this lizard has the genus's typical broad triangular head, stout body, relatively short limbs, short tail and blue tongue. Members of southern populations are silvery-grey to yellow-brown above, with a series of dark-brown to black cross-bars extending from the forelimb to the tail. A wide black streak extends from the eye to above the ear. Northern populations are a rich brown with paler bars which are often obscure or broken into mottling; on the sides there are large alternate black and cream to orange bars and there is seldom a dark streak from the eye. Occurring in northern Western Australia, northern areas of the Northern Territory, much of Queensland, New South Wales and Victoria, and south-eastern South Australia, it is found in a wide variety of habitats, where it shelters in burrows, ground debris and crevices. It feeds on insects, snails, carrion and vegetable matter and females produce five to 25 young in a litter.

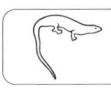

Shingleback (*Trachydosaurus rugosus*) 30–40 cm

One of the darker forms of T. rugosus.

With its large pine-cone-like scales, and short, rounded tail, this lizard can scarcely be mistaken for anything else. It has a large angular head, stout body and a blue tongue. The colour ranges from red-brown to black, with or without cream to yellow spots. In some populations the light colour predominates and the dark colour forms the spots. It occurs in southern Western Australia, southern South Australia, northern and western Victoria, New South Wales (except the coast), and southern Queensland (except the coast). It is found in a wide variety of habitats, where it shelters under leaf litter, ground debris or timber. A slow-moving lizard, it feeds upon insects, carrion and vegetable matter. Females produce one to three large young in a litter.

Glossary

Anal. Referring to the area of the body at and immediately surrounding the anus or cloaca.

Arboreal. Living in trees or shrubs.

Arthropod. Any member of largest invertebrate group (phylum Arthropoda) which includes insects, spiders, centipedes and millipedes.

Autotomy. The voluntary casting off of a part of the body. Certain lizards shed their tails by this process, usually as a defence mechanism.

Carapace. The upper half of the shell of a tortoise or turtle.

Carnivorous. Feeding on animal tissues.

Costal. In turtles and tortoises, the name given to the large shields or scales which form two rows on the carapace, on either side of the row of vertebrals.

Crepuscular. Active mainly at dusk or in the early morning.

Distal. Located furthest away from the point of attachment.

Diurnal. Active during the day.

Dorsal. Pertaining to the back or upper surfaces of the body.

Ectothermic. Descriptive of those animals, such as reptiles, which regulate body temperature using outside heat sources.

Exfoliations. Split sheets or slabs of flaking rock (caused by weathering).

Fossorial. Burrowing (and normally living mostly beneath the surface of the soil).

Herbivorous. Feeding on plant material.

Imbricate. In reptiles, descriptive of body scales that overlap.

Invertebrate. Lacking a backbone. The invertebrates are the lower animals such as worms, snails, insects, spiders etc.

Keeled. Of reptile scales, having a ridge or ridges.

Labial. In reptiles, the name given to the scales along the lips.

Lateral. Pertaining to the sides.

Loreal. In reptiles, the name given to a scale on the side of the snout between those of the nostril (the nasals) and the eye (the preoculars).

Nape. The back of the neck.

Nocturnal. Active during the night.

Ocellus. (Plural ocelli.) Circular, eye-like, pigmented spots made up of a central spot and one or more concentric rings, which occur as a pattern on some reptiles.

Parthenogenesis. Reproduction by females without fertilization by males.

Plastron. The lower half of the shell of a tortoise or turtle, protecting the underparts.

Prehensile. Of reptile tails, adapted for grasping or wrapping around.

Proximal. Located closest to the body or to the point of attachment.

Rostral. The scale situated at the front of the snout, between the scales that border the sides of the upper lip.

Rugose. Wrinkled or corrugated.

Shields. The large scales on the shell of a tortoise or turtle. (Sometimes called scutes.)

Spur. In pythons, a vestigial hindlimb which has been reduced to a sharp-pointed, rigid projection on either side of the vent.
Termitarium. (Plural termitaria.) The nest of a termite colony.
Terrestrial. Living on the ground.
Trilobed. With three lobes.
Tubercle. A small knob-like protuberance on the skin.
Valvular. Equipped with a valve-like structure, allowing a flow in one direction only.
Vent. The external entrance of the cloaca or anus.
Ventral. Pertaining to the underparts (of an animal).
Vertebral. In turtles and tortoises, the name given to each of the shields or scales down the middle of the carapace. As an adjective, descriptive of the area over the spinal column.

Further Reading

Cann, J. 1978. *Tortoises of Australia*, Angus & Robertson, Sydney.
Cogger, H.G. 1992. *Reptiles and Amphibians of Australia.* Reed Books, Sydney.
Ehmann, H. 1992. *Encyclopedia of Australian Animals: Reptiles.* Angus & Robertson, Sydney.
Gow, G.F. 1989. *Graeme Gow's Complete Guide to Australian Snakes.* Angus & Robertson, Sydney.
Greer, A.E. 1989. *The Biology and Evolution of Australian Lizards*, Surrey Beatty & Sons, Sydney.
Jenkins, R. & Bartell, R. 1980. *A Field Guide to Reptiles of the Australian High Country*, Inkata Press, Melbourne.
Mirtschin, P. & Davis, R. 1992. *Snakes of Australia, Dangerous & Harmless.* Hill of Content, Melbourne.
Schmida, G. 1985. *The Cold-blooded Australians.* Doubleday, Sydney.
Shine, R. 1991. *Australian Snakes: A Natural History.* Reed Books, Sydney.
Storr, G.M., Smith, L.A. & Johnstone, R.E. 1981. *Lizards of Western Australia I: Skinks.* 1983. *Lizards of Western Australia II: Dragons and Monitors.* 1986. *Snakes of Western Australia.* 1990. *Lizards of Western Australia III: Geckos and Pygopods.* Western Australian Museum, Perth.
Swan, G. 1990. *A Field Guide to the Snakes and Lizards of New South Wales.* Three Sisters Productions, Winmalee, New South Wales.
Webb, G. & Manolis, C. 1989. *Crocodiles of Australia.* Reed Books, Sydney.
Weigel, J. 1990. *The Australian Reptile Park's Guide to the Snakes of South-east Australia*, Australian Reptile Park, Gosford, New South Wales.
Wilson, S. & Knowles, D. 1988. *Australia's Reptiles: A Photographic Reference to the Terrestrial Reptiles of Australia.* Collins, Sydney.

Index

Acalyptophis peronii 64
Acanthophis antarcticus 30
　A. praelongus 31
　A. pyrrhus 31
Aclys concinna 91
Acrochordus arafurae 22
　A. granulatus 23
Adder. see Death Adder
Aipysurus duboisii 64
　A. laevis 65
Amphibolurus muricatus 103
　A. nobbi 103
　A. norrisi 104
Anomalopus verreauxii 113
Antaresia childreni 17
　A. maculosa 17
　A. perthensis 18
　A. stimsoni 18
Aprasia inaurita 91
Aspidites melanocephalus 14
　A. ramsayi 15
Astrotia stokesii 65
Austrelaps labialis 32
　A. ramsayi 32
　A. superbus 33
Banded Snake
　De Vis's 37
　Southern Desert 57
　Stephens' 43
Bandy-bandy 63
Bardick 39
Bassiana platynotum 135
Black Snake
　Blue-bellied 49
　Red-bellied 50
Blind Snake
　Blackish 13
　Northern 12
　Prong-snouted 12
　Proximus 13
　Southern 11
Bluetongue
　Blotched 138
　Centralian 137
　Common 138
　Common Slender 119
Boiga fusca 24
　B. irregularis 25
Brown Snake
　Eastern 53
　Ingram's 52
　Ringed 54
　Speckled 52
　Western 51
Brown Tree-snake. see Tree-snake
Burrowing Snake
　Black-naped 58
　Black-striped 58
Cacophis harriettae 33
　C. krefftii 34
　C. squamulosus 34
Caretta caretta 73
Carettochelys insculpta 76
Carlia jarnoldae 114
　C. munda 114
　C. tetradactyla 115

Carpet Python 20
　Centralian 19
　West Australian 21
Chelodina expansa 77
　C. longicollis 77
　C. rugosa 78
Chelonia mydas 74
Chlamydosaurus kingii 104
Christinus marmoratus 80
Cool-skink
　Red-throated 135
　Tussock 136
Copperhead
　Highland 32
　Lowland 33
　Pygmy 32
Coppertail 118
Coral Snake 57
Crenadactylus ocellatus 81
Crevice-dragon
　Tawny 105
Crevice-skink
　South-western 123
　Tree 124
Crocodile
　Estuarine 71
　Freshwater 72
Crocodylus johnstoni 72
　C. porosus 71
Crowned Snake 37
　Dwarf 34
　Golden- 34
　White- 33
Cryptoblepharus plagiocephalus 115
　C. virgatus 116
Ctenophorus decresii 105
　C. fordi 106
　C. isolepis 106
　C. nuchalis 106
Ctenotus
　Southern Mallee 116
　Leopard 117
　Robust 117
　Barred Wedge-snout 118
Ctenotus atlas 116
　C. pantherinus 117
　C. robustus 117
　C. schomburgkii 118
　C. taeniolatus 118
Cyclodomorphus branchialis 119
　C. casuarinae 119
Death Adder
　Desert 31
　Northern 31
　Southern 30
Delma
　Excitable 93
　Many-lined 92
　Sharp-snouted 92
Delma impar 92
　D. nasuta 92
　D. tincta 93
Demansia atra 35
　D. olivacea 35
　D. psammophis 36
　D. torquata 36
Dendrelaphis punctulata 26
Denisonia devisi 37

Dermochelys coriacea 75
Desert-skink
　Unadorned 122
Devil Thorny 109
Diplodactylus ciliaris 82
　D. elderi 82
　D. pulcher 83
　D. vittatus 83
　D. williamsi 84
Diporiphora australis 107
　D. winneckei 107
Dragon
　Blotch-tailed Earless 111
　Canegrass Two-line 107
　Central Bearded 111
　Crevice. see Crevice-dragon
　Eastern Bearded 110
　Eastern Two-line 107
　Eastern Water 109
　Ground. see Ground-dragon
　Heath. see Heath-dragon
　Lined Earless 112
　Nobbi, Lashtail, Thorny Devil.)
　Sand. see Sand-dragon
　Southern Angle-headed 108
　Western Bearded 110
Dragon (see also Frilled Lizard, Jacky Lizard, Nobbi, Lashtail, Thorny Devil.)
Drysdalia coronata 37
　D. coronoides 38
　D. mastersii 39
Dtella
　Variegated 84
Dugite 50
Dwarf Skink
　Common 134
Earless Skink
　Three-toed 129
　Lowlands 129
Echiopsis curta 39
Egernia cunninghami 121
　E. frerei 120
　E. inornata 122
　E. kingii 122
　E. major 123
　E. napoleonis 123
　E. stokesii 124
　E. striolata 124
　E. whitii 125
Elseya dentata 78
Emydocephalus annulatus 66
Emydura krefftii 79
　E. macquarii 79
Enhydris polylepis 29
Ephalophis greyi 67
Eremiascincus richardsonii 125
Eretmochelys imbricata 75
Eulamprus heatwolei 126
　E. kosciuskoi 126
　E. quoyii 127
　E. tenuis 127
File Snake
　Arafura 22
　Little 23

Firetail-skink
  Lined 135
Fordonia leucobalia 29
Forest-skink
  Bar-sided 127
  Prickly 128
Furina diadema 40
  F. ornata 41
Garden Sun-skink. see
  Sun-skink, Garden
Gecko
  Bynoe's Prickly 85
  Clawless 81
  Eastern Spiny-tailed 84
  Eastern Stone 83
  Fine-faced 83
  Jewelled 82
  Leaf-tailed 89
  Marbled Southern 80
  Marbled Velvet 87
  Northern Spiny-tailed 82
  Rough Knob-tailed 86
  Southern leaf-tailed 88
  Thick-tailed 89
  Three-lined Knob-tailed 86
  Zig-zag Velvet 88
Gecko (see also Dtella)
Gehyra variegata 84
Glaphyromorphus isolepis 128
Gnypetoscincus queenslandiae 128
Goanna. see Monitor
Ground-dragon
  Central Netted 106
Heath-dragon
  Mountain 112
Hemiaspis damelii 41
  H. signata 42
Hemiergis decresiensis 129
  H. peronii 129
Hemisphaeriodon gerrardii 130
Heteronotia binoei 85
Hooded Snake
  Spectacled 62
Hoplocephalus bitorquatus 42
  H. bungaroides 43
  H. stephensii 43
Hydrelaps darwiniensis 67
Hydrophis elegans 68
  H. ornatus 68
Hypsilurus spinipes 108
Keelback 28
Krait. see Sea Krait
Lampropholis delicata 131
  L. guichenoti 131
Lapemis hardwickii 69
Lashtail
  Gilbert's 108
  Mallee Heath 104
Laticauda colubrina 70
Legless Lizards (family
  Pygopodidae – see
  Delma, Lizard, Scaly-
  foot, Slider, Snake-
  lizard, Worm-lizard.)
Lerista bipes 132

  L. bougainvillii 132
  L. muelleri 133
  L. planiventralis 133
Lialis burtonis 90
Liasis fuscus 16
  L. olivaceus 16
Lizard
  Frilled 104
  Jacky 103
  Javelin 91
  Snake-. see Snake-
    lizard
  Worm-. see Worm-
    lizard
Lophognathus gilberti 108
Mangrove Snake. see
  Snake, Mangrove
Meadow-skink, Alpine 126
Menetia greyii 134
Moloch horridus 111
Monitor
  Black-tailed 100
  Heath 101
  Lace 102
  Mertens' Water 99
  Ocellate Ridge-tailed 96
  Pygmy Mulga 96
  Rusty Desert 97
  Sand 99
  Short-tailed Pygmy 97
  Spencer's 101
  Storr's 102
  Yellow-spotted 100
Monitor (see also
  Perentie.)
Morelia amethistina 19
  M. bredli 19
  M. oenpelliensis 20
  M. spilota imbricata 21
  M. spilota spilota 21
  M. spilota variegata 20
  M. viridis 15
Morethia
  South-eastern 134
Morethia boulengeri 134
  M. ruficauda 135
Mulch-slider
  Wood 133
Mulga Snake. see Snake,
  Mulga
Mullet, Land 123
Nephrurus asper 86
  N. levis 86
  N. milii 89
Nobbi 103
Notechis ater ater 44
  N. ater niger 44
  N. ater humphreysi 45
  N. scutatus scutatus 45
  N. scutatus occidentalis 46
Oedura marmorata 87
  O. rhombifer 88
Oxyuranus microlepidotus 46
  O. scutellatus 47
Paradelma orientalis 91
Pelamis platurus 69
Perentie 98
Phyllurus platurus 88

Physignathus lesueurii 109
Pletholax gracilis 94
Pogona barbata 110
  P. minor 109
  P. vitticeps 110
Pseudechis australis 47
  P. butleri 48
  P. colletti 48
  P. guttatus 49
  P. porphyriacus 50
Pseudemoia entrecasteauxii 136
Pseudonaja affinis 50
  P. guttata 52
  P. ingrami 52
  P. modesta 54
  P. nuchalis 51
  P. textilis 53
Pygopus lepidopodus 94
  P. nigriceps 95
Python
  Amethystine 19
  Black-headed 14
  Carpet 20
  Centralian Carpet 19
  Children's 17
  Diamond 21
  Eastern Small-blotched 17
  Green 15
  Large-blotched 18
  Oenpelli Rock 20
  Olive 16
  Pygmy 18
  Water 16
  West Australian Carpet 21
Python (see also Carpet
  Python, Woma.)
Rainbow-skink
  Lined 114
  Shaded-litter 114
  Southern 115
Ramphotyphlops australis 11
  R. bituberculatus 12
  R. diversus 12
  R. nigrescens 13
  R. proximus 13
Rhinoplocephalus boschmai 55
  R. nigrescens 55
  R. nigrostriatus 56
Saiphos equalis 136
Saltuarius swaini 89
Sand-dragon
  Mallee 105
  Military 106
Sandslider
  North-western 132
Sand-swimmer
  Broad-banded 125
Saproscincus mustelina 137
Scaly-foot
  Brigalow 93
  Hooded 94
  Southern 94
Sea Krait
  Wide-faced 70
Sea-snake
  Elegant 68
  Horned 64

143

Black-ringed Mangrove 67
North-western Mangrove 67
Olive 65
Ornate 68
Reef Shallows 64
Spine-bellied 69
Stokes' 65
Turtle-headed 66
Yellow-bellied 69
Shingleback 139
Shinning-skink
Callose-palmed 115
Cream-striped 116
Shovel-nosed Snake
Narrow-banded 59
North-western 56
Southern 59
Simoselaps approximans 56
S. australis 57
S. bertholdi 57
S. bimaculatus 58
S. calonotus 58
S. fasciolatus 59
S. semifasciatus 59
Skink
Alpine Meadow-. see Meadow-skink.
Common Dwarf Skink 134
Cool. see Cool-skink
Crevice. see Crevice-skink
Cunningham's 121
Desert. see Desert-skink
Firetail. see Firetail-skink
Forest. see Forest-skink
Gidgee 124
King's 122
Lowlands Earless 129
Major 120
Northern Bar-lipped 128
Pink-tongued 130
Rainbow. see Rainbow-skink
She-oak 119
Shinning. see Shinning-skink
Sun. see Sunskink
Three-toed Earless 129
Water. see Water-skink
Weasel 137
White's 125
Worm. see Worm-skink
Yellow-bellied Three-toed 136
Skink (see also Bar-lipped Skink, Bluetongue, Coppertail, Ctenotus, Dwarf Skink, Earless Skink, Morethia, Mulch-slider, Mullet, Sandslider, Shingleback, Slider.)
Slender Bluetongue. see Bluetongue, Slender
Slider

Keeled 133
Mulch. see Mulch-slider
Sand. see Sandslider
Slender 94 (family Pygopodidae)
South-eastern 132
Snake
Banded. see Banded Snake
Black. see Black Snake
Black-striped 56
Blind. see Blind Snake
Broad-headed 43
Brown. see Brown Snake
Burrowing. see Burrowing Snake
Collett's 48
Coral 57
Crowned. see Crowned Snake
Curl 62
Dwarf Crowned 34
Eastern Carpentaria 55
File. see File Snake
Golden-crowned 34
Grey 41
Hooded. see Hooded Snake
Macleay's Water 29
Mallee Black-backed 61
Marsh 42
Masters' 39
Monk 60
Mulga 48
Orange-naped 41
Pale-headed 42
Red-naped 40
Rough-scaled 63
Sea. see Sea-snake
Shovel-nosed. see Shovel-nosed Snake
Slate-grey 27
Small-eyed 55
Spotted 61
Spotted Mulga 48
Tiger. see Tiger Snake
Tree. see Tree-snake
Whip. see Whipsnake
White-bellied Mangrove 29
White-crowned 33
White-lipped 38
Snake-lizard
Burton's 90
Stegonotus cucullatus 27
Sunskink
Dark-flecked Garden 131
Pale-flecked Garden 131
Suta flagellum 60
S. monachus 60
S. nigriceps 61
S. punctata 61
S. spectabilis 62
S. suta 62
Taipan 47
Inland 46
Thorny Devil 111
Three-toed Skink. see

Skink, Yellow-bellied Three-toed
Tiger Snake
Eastern 45
Krefft's 44
Peninsular 44
Tasmanian 45
Western 46
Tiliqua multifasciata 137
T. nigrolutea 138
T. scincoides 138
Tortoise
Murray Short-necked 79
Trachydosaurus rugosus 139
Tree-snake
Eastern Brown 25
Green 26
Northern Brown 24
Tropidechis carinatus 63
Tropidonophis mairii 28
Turtle
Broad-shelled Snake-necked 77
Eastern Snake-necked 77
Green 74
Hawksbill 75
Krefft's River 79
Leathery 75
Loggerhead 73
Northern Snake-necked 78
Northern Snapping 78
Pitted-shelled 76
Tympanocryptis cephalus 111
T. diemensis 112
T. lineata 112
Varanus acanthurus 96
V. brevicauda 97
V. eremius 97
V. giganteus 98
V. gilleni 98
V. gouldii 99
V. mertensi 99
V. panoptes 100
V. rosenbergi 101
V. spenceri 101
V. storri 102
V. tristis 100
V. varius 102
Vermicella annulata 63
Water-skink
Eastern 127
Warm-temperate 126
Water Snake. see Snake, Water
Wedge-snout Ctenotus. see Ctenotus, Wedge-snout
Whipsnake
Black 35
Collared 36
Little 60
Olive 35
Yellow-faced 36
Woma 15
Worm-lizard
Mallee 91
Worm-skink
Three-clawed 113